DRINKING WATER AND HEALTH

Volume 9:
Selected Issues in Risk Assessment

Safe Drinking Water Committee
Board on Environmental Studies
 and Toxicology
Commission on Life Sciences
National Research Council

NATIONAL ACADEMY PRESS
Washington, D.C. 1989

NATIONAL ACADEMY PRESS, 2101 Constitution Ave., NW, Washington, DC 20418

List of Participants

SUBCOMMITTEE ON DNA ADDUCTS

DAVID J. BRUSICK, Hazleton Laboratories America, Inc., Vienna, Virginia, *Chairman*

GAIL T. ARCE, Haskell Laboratory for Toxicology and Industrial Medicine, E. I. du Pont de Nemours & Co., Newark, Delaware

JOHN C. BAILAR, McGill University School of Medicine, Montreal, Quebec, Canada

RAMESH C. GUPTA, Baylor College of Medicine, Houston, Texas

ROBIN HERBERT, Mount Sinai Medical Center, New York, New York

PAUL H. M. LOHMAN, Sylvius Laboratory, State University of Leiden, Leiden, The Netherlands

CAROL W. MOORE, CUNY Medical School, The City College of New York, New York, New York

ROBERT F. MURRAY, Howard University College of Medicine, Washington, D.C.

MIRIAM C. POIRIER, National Cancer Institute, Bethesda, Maryland

GARY A. SEGA, Biology Division, Oak Ridge National Laboratory, Oak Ridge, Tennessee

RICHARD B. SETLOW, Biology Department, Brookhaven National Laboratory, Upton, New York

JAMES A. SWENBERG, Chemical Industry Institute of Toxicology, Research Triangle Park, North Carolina

Advisers and Contributors

ROGER W. GIESE, College of Pharmacy and Allied Health Professions, Northeastern University, Boston, Massachusetts

FRANK C. RICHARDSON, Chemical Industry Institute of Toxicology, Research Triangle Park, North Carolina

SUBCOMMITTEE ON MIXTURES

RONALD WYZGA, Electric Power Research Institute, Palo Alto, California, *Chairman*

JULIAN B. ANDELMAN, University of Pittsburgh Graduate School of Public Health, Pittsburgh, Pennsylvania

W. HANSBROUGH CARTER, JR., Virginia Commonwealth University, Richmond, Virginia

NANCY R. KIM, New York State Department of Health, Albany, New York

SHELDON D. MURPHY, University of Washington, Seattle, Washington

BERNARD WEISS, University of Rochester School of Medicine and Dentistry, Rochester, New York

RAYMOND S. H. YANG, National Institute of Environmental Health Sciences/National Toxicology Program, Research Triangle Park, North Carolina

SAFE DRINKING WATER COMMITTEE

DAVID J. JOLLOW, Medical University of South Carolina, Charleston, South Carolina, *Chairman*

DAVID E. BICE, Lovelace Inhalation Toxicology Research Institute, Albuquerque, New Mexico

JOSEPH F. BORZELLECA, Virginia Commonwealth University, Richmond, Virginia

DAVID J. BRUSICK, Hazleton Laboratories America, Inc., Vienna, Virginia

EDWARD J. CALABRESE, North East Regional Environmental Public Health Center, University of Massachusetts, Amherst, Massachusetts

J. DONALD JOHNSON, School of Public Health, University of North Carolina, Chapel Hill, North Carolina

RONALD WYZGA, Electric Power Research Institute, Palo Alto, California

National Research Council Staff

ANDREW M. POPE, *Project Director*

MARVIN SCHNEIDERMAN, *Principal Staff Scientist*

LESLYE B. WAKEFIELD, *Project Coordinator* (until September 1988)

ANNE M. SPRAGUE, *Research Assistant*

NORMAN GROSSBLATT, *Editor*

Sponsoring Agency

KRISHAN KHANNA, Office of Drinking Water, U.S. Environmental Protection Agency, Washington, D.C., *Technical Manager*

BRUCE MINTZ, Office of Drinking Water, U.S. Environmental Protection Agency, Washington, D.C., *Project Officer*

BOARD ON ENVIRONMENTAL STUDIES AND TOXICOLOGY

GILBERT S. OMENN, School of Public Health and Community Medicine, University of Washington, Seattle, Washington, *Chairman*

FREDERICK R. ANDERSON, Washington College of Law, American University, Washington, D.C.

JOHN C. BAILAR, McGill University School of Medicine, Montreal, Quebec, Canada

DAVID BATES, University of British Columbia Health Science Center, Vancouver, British Columbia, Canada

JOANNA BURGER, Department of Biological Sciences, Rutgers University, Piscataway, New Jersey

RICHARD A. CONWAY, Department of Engineering, Union Carbide Corporation, South Charleston, West Virginia

WILLIAM E. COOPER, Department of Zoology, Michigan State University, East Lansing, Michigan

SHELDON K. FRIEDLANDER, Department of Chemical Engineering, University of California, Los Angeles, California

BERNARD D. GOLDSTEIN, University of Medicine and Dentistry of New Jersey-Robert Wood Johnson Medical School, Piscataway, New Jersey

DONALD R. MATTISON, University of Arkansas for Medical Sciences, Little Rock, Arkansas

DUNCAN PATTEN, Arizona State University Center for Environmental Studies, Tempe, Arizona

EMIL A. PFITZER, Department of Toxicology and Pathology, Hoffmann-La Roche Inc., Nutley, New Jersey

PAUL RISSER, University of New Mexico, Albuquerque, New Mexico

WILLIAM H. RODGERS, University of Washington School of Law, Seattle, Washington

F. SHERWOOD ROWLAND, Department of Chemistry, University of California, Irvine, California

LIANE B. RUSSELL, Biology Division, Oak Ridge National Laboratory, Oak Ridge, Tennessee

MILTON RUSSELL, Energy Division, Oak Ridge National Laboratory, Oak Ridge, Tennessee

ELLEN K. SILBERGELD, Environmental Defense Fund, Washington, D.C.

I. GLENN SIPES, University of Arizona College of Pharmacy, Tucson, Arizona

COMMISSION ON LIFE SCIENCES

BRUCE M. ALBERTS, Department of Biochemistry and Biophysics, University of California, San Francisco, California, *Chairman*

PERRY L. ADKISSON, Chancellor, Texas A&M University System, College Station, Texas

FRANCISCO J. AYALA, Department of Ecology and Evolutionary Biology, University of California, Irvine, California

J. MICHAEL BISHOP, The G. W. Hooper Research Foundation, University of California Medical Center, San Francisco, California

FREEMAN J. DYSON, School of Natural Sciences, The Institute for Advanced Study, Princeton, New Jersey

NINA V. FEDOROFF, Department of Embryology, Carnegie Institution of Washington, Baltimore, Maryland

RALPH W. F. HARDY, Boyce Thompson Institute for Plant Research, Ithaca, New York

RICHARD J. HAVEL, Cardiovascular Research Institute, University of California School of Medicine, San Franciso, California

LEROY E. HOOD, Division of Biology, California Institute of Technology, Pasadena, California

DONALD F. HORNIG, Interdisciplinary Programs in Health, Harvard School of Public Health, Boston, Massachusetts

ERNEST G. JAWORSKI, Division of Biological Sciences, Monsanto Company, St. Louis, Missouri

SIMON A. LEVIN, Ecosystems Research Center, Cornell University, Ithaca, New York

HAROLD A. MOONEY, Department of Biological Sciences, Stanford University, Stanford, California

STEPHEN P. PAKES, Southwestern Medical School, University of Texas, Dallas, Texas

JOSEPH E. RALL, Intramural Research, National Institutes of Health, Bethesda, Maryland

RICHARD D. REMINGTON, Academic Affairs, University of Iowa, Iowa City, Iowa

PAUL G. RISSER, University of New Mexico, Alburquerque, New Mexico

RICHARD B. SETLOW, Biology Department, Brookhaven National Laboratory, Upton, New York

TORSTEN N. WIESEL, Laboratory of Neurobiology, Rockefeller University, New York, New York

National Research Council Staff

JOHN E. BURRIS, *Executive Director*, Commission on Life Sciences
DEVRA L. DAVIS, *Director*, Board on Environmental Studies and
  Toxicology
JAMES J. REISA, *Associate Director*
RICHARD D. THOMAS, *Director*, Toxicology and Epidemiology Program
LEE R. PAULSON, *Manager*, Toxicology Information Center
JACQUELINE K. PRINCE, *Administrative Associate*
JEANETTE SPOON, *Administrative Assistant*

Preface

The 1974 Safe Drinking Water Act (U.S. Public Law 93-523) authorized the U.S. Environmental Protection Agency (EPA) to establish federal standards to protect the public from harmful contaminants of drinking water. The law also provided for the establishment of a joint national-state system to ensure compliance with the standards and to protect underground water sources from contamination. Section 1412(e) of the act and its amendments (42 U.S. Code, Subpart 300f et seq., 1977) mandated that the National Research Council (NRC) conduct studies to identify adverse health effects associated with contaminants in drinking water, to identify relevant research needs, and to make recommendations regarding such research. Amendments to the law in 1977 requested revisions of the NRC studies to report "new information which had become available since the NRC's most recent report, and every two years thereafter."

This is the ninth volume in the series *Drinking Water and Health* issued by the Safe Drinking Water Committee of the Board on Environmental Studies and Toxicology of the NRC. Each volume has reviewed toxicologic data and assessed risks associated with specific drinking water contaminants. This volume focuses on two important current topics: the first part examines the possible uses of DNA adducts (addition products) in risk assessment, and the second part examines the issue of multiple toxic chemicals in drinking water and the assessment of their health risks. A comprehensive index to all nine volumes of the *Drinking Water and Health* series is also provided in this volume.

As described in Part 1, studies of DNA have been rapidly refined and developed in the past few years. The ability to detect ever-smaller molecular

alterations of DNA provides important opportunities for estimating and re-
ducing public health risks associated with drinking water contaminants, foods,
and workplace chemicals that bind to DNA to form adducts. In addition,
protein adducts found in easily accessible body fluids sometimes reflect
potential DNA-adduct formation.

In recognition of the potential of these recent advances for protecting
human health, the EPA's Office of Drinking Water asked the NRC's Safe
Drinking Water Committee to convene a small group of experts in DNA-
adduct research to review developments in the field (with special attention
to possible uses of DNA adducts in risk assessment). EPA was especially
interested in whether current techniques could confirm exposure or signal
tumor initiation. The group was also to point out gaps in research and suggest
priorities for additional research.

The introduction of new methods to measure DNA adducts and protein
adducts has already made some types of direct human population monitoring
technically and economically feasible. That application of DNA technology
might permit epidemiologic confirmation of reported human exposures and
offer opportunities to validate extrapolation of data from animal bioassays.
Participants in several recent meetings on research in this field have agreed
that the new methods can have a marked impact on the biologic bases for
estimating the risks associated with human exposure to several important
classes of environmental pollutants. Use of the technology in risk assessment
will depend on an understanding of the mechanistic relationships between
DNA alterations and the ultimate expression of toxic effects. Recent developments in the study of DNA binding and protein binding have provided a
useful tool to begin to acquire that understanding, but additional information,
such as clarification of the role of background or baseline adducts that are
continually formed in animals and humans, will be needed for full use of
the techniques.

New methods of measuring DNA adducts are useful for several other
reasons:

• They have permitted increasingly refined measurements of genetic material.
• They meet requirements of high intrinsic sensitivity and specificity at
exposures approaching those in occupational and environmental settings.
• They seem in many instances to be relatively inexpensive, fast, and
reproducible.
• They can be applied to readily available samples of such body fluids as
blood, urine, and semen and to small samples of cells, such as of buccal
mucosa and skin.

Analytic instrumentation and relatively noninvasive methods for measuring
DNA adducts and protein adducts could eventually lead to improved under-

standing of the mechanisms of carcinogenesis, mutagenesis, and other health effects of exposure to DNA damaging agents. The Executive Summary of Part 1 summarizes the findings of the subcommittee. Chapter 1 describes where and how DNA adducts are formed and repaired; what is known of their relationship to protein adducts and to exposure to, and toxic effects of, contaminants; and some differences in adduct formation between humans and laboratory animals. The uses and limitations of current techniques for detecting DNA adducts and protein adducts and the outlook for the application of the techniques in toxicity testing, biologic monitoring, and epidemiology are described in Chapter 2. Chapter 3 presents the subcommittee's conclusions and recommendations. An appendix characterizes selected contaminants found in drinking water and identifies those known to bind to DNA and form adducts, and a glossary defines terms.

Part 2 of this volume addresses mixtures of toxic chemicals. The toxicity of chemicals is traditionally studied in terms of the effects of exposure to single toxic substances, rather than mixtures of substances. Regulatory agencies have used results of studies of single toxicants to form procedures for regulating exposure. But predicting effects of mixtures solely from knowledge of effects of their components can be erroneous. Some agents interact when combined to produce biologic responses different from those expected, and interactions and the magnitude of responses might not be considered properly. Many components of drinking water produce similar biologic effects. For example, the volatile, halogenated hydrocarbons are known to form common metabolites in mammalian systems. Other components inhibit enzymes in common, follow common metabolic pathways, or have common mechanisms of action in target organs.

EPA's Office of Drinking Water asked the Safe Drinking Water Committee to convene a workshop to address the issue of mixtures of chemicals in drinking water and explore the improvement of methods for assessing the risk associated with chronic, low-level exposure to such mixtures. In light of the apparently common characteristics of some of the many chemicals in drinking water, EPA was particularly interested in the possibility of grouping some drinking water constituents for combined risk assessment. As part of the workshop, the Safe Drinking Water Committee's Subcommittee on Mixtures reviewed the 1988 NRC report *Complex Mixtures* and related literature. The subcommittee then suggested to regulators that some drinking water contaminants, such as the volatile organic chemicals or organophosphorus and carbamate insecticides, could be grouped for combined risk assessment. Furthermore, the subcommittee suggested that even the risks associated with exposures to unlike chemicals that produce a wide variety of health effects might be weighed and combined.

On behalf of the members of our two subcommittees, we would like to express our gratitude to the NRC staff members who assisted in these projects.

Leslye Wakefield served as project coordinator; her tireless efforts greatly aided both subcommittees in completing their work. We thank Lee Paulson, manager of the Toxicology Information Center, and Anne Sprague, who completed the final manuscript. Richard Thomas directed the NRC Safe Drinking Water programs for several years and was the initial director of this study. He also served as technical adviser on the DNA adducts portion of this report. Andrew Pope served as project director of the Mixtures portion of the report. The subcommittee also acknowledges the efforts of Norman Grossblatt, who edited the report, and those of Alison Kamat, Linda Poore, Erik Hobbie, and Bernidean Williams, who assisted in extensive searching of the scientific literature and in reference verification. We also acknowledge the help of Robin Bowers in preparing manuscripts and Erin Schneider in providing general secretarial support.

We especially thank Marvin Schneiderman, whose expertise and invaluable assistance were necessary to complete the second part of this report on chemical mixtures. We are grateful to Frank Richardson, who served as an adviser to our subcommittees; to Safe Drinking Water Committee Chairman David Jollow for his valuable participation; and to Devra Davis and Alvin Lazen for their creative insight and guidance. Our special thanks go to John Bailar for guiding this report to completion. Finally, we thank our colleagues on the two subcommittees for their contributions to this report.

DAVID BRUSICK, *Chairman*
Subcommittee on DNA Adducts

RONALD WYZGA, *Chairman*
Subcommittee on Mixtures

Contents

xiii

PART III: CUMULATIVE INDEX

DRINKING
WATER
AND
HEALTH

PART I
DNA Adducts

Executive Summary

DNA (deoxyribonucleic acid), a complex substance composed of polymers of small molecules called nucleotides, is present in the nuclei of all cells and is the carrier of genetic information. Recent advances in methodology and instrumentation permit detection and measurement of alterations within the individual nucleotides of DNA. In one important kind of DNA alteration, exogenous and xenobiotic materials bind to nucleotides within DNA to form addition products, or adducts. Radiolabeling, immunochemical, and physical methods can detect adducts at concentrations as low as one in 10^9-10^{10} nonadducted nucleotides.

DNA adducts can form in many tissues, but they are not necessarily stable. They can decompose spontaneously, and a number of them can be removed by enzymatic repair systems at varying rates. The toxicologic significance of the presence of background adducts or even of induced adducts at persistently high concentrations is unknown. There is reason to regard some DNA adducts as markers of exposure to specific toxicants. Research must distinguish those from "background" adducts, and their long-term impact on organisms needs to be assessed. In addition, protein adducts, such as those found in the hemoglobin of red blood cells and in sperm protamine, are apparently stable for the lifetime of the cell and are thus good indicators of recent exposure. Protein adducts should be considered in the estimation of genetic or carcinogenic risk whenever they can be correlated with DNA binding, even though they themselves may play no putative mechanistic role.

Interpretation of the presence of DNA adducts caused by even a single known carcinogen is highly complex. For example, whether an adduct is crucial to tumor initiation depends heavily on the chemical and toxicologic

3

properties of the compound of concern, on the metabolic system of the host organism, and on the site of the adduct within a specific nucleotide, gene, and target tissue. Current evidence suggests associations between the occurrence of DNA adducts formed by specific compounds and various types of toxic effects (usually tissue-specific), such as mutation, cancer, and developmental effects, although in no case has a specific DNA adduct been firmly established as the cause of a tumor or mutation in a mammalian system in vivo. The presence of DNA adducts in a tissue does not necessarily establish risk, but is often an indication that additional testing for toxicity should be done.

The technology for detecting and measuring adducts is qualitatively useful in the toxicologic evaluation of specific chemical contaminants of drinking water, but the only quantitative application that has been validated is in the assessment of exposure. Future quantitative applications might be in dosimetry, pharmacokinetics, hazard identification, and risk assessment; DNA-adduct technology might also be applied eventually to large-scale human monitoring studies.

The technology and the database have deficiencies that must be resolved before DNA-adduct detection, identification, and measurement can be viewed as routine components of general toxicity testing. With the newest technology, many adducts of unidentified chemical nature or unknown stability can be detected and counted, but the biologic impact of their presence cannot be established without more information. For example, the relationship between adduct formation and tumorigenesis varies with the chemical compound, the specific target tissue, the organism's exposure history, the duration and time of exposure, and so on. Large-scale DNA-adduct dosimetry studies in humans are now becoming possible, but they will be subject to much more variability because of population heterogeneity and thus will have many more problems than corresponding dosimetry studies in animal models. Even with population studies to identify variations due to age, sex, race, and interference factors, it will be much more difficult to predict extent of exposure or cancer risk based on DNA-adduct formation in humans than in homogeneous laboratory animals.

The real advantage of gaining information about DNA adducts lies in the addition of this knowledge to the comprehensive database needed for assessing the risk of exposure to chemicals with identified toxicity. This information could be used in individual risk assessment to confirm suspected exposures, improve estimates of target tissue dose, and reveal metabolic activation and detoxification rates for specific carcinogen–DNA-adduct formation. In general risk assessment, it could be extremely valuable in estimating dosimetry and systemic distribution and in establishing possible target tissues or organs and the potential for irreversible toxicity, such as cancer, mutation, or developmental effects. The use of DNA adducts as molecular dosimeters might make possible the study of differences in absorption, distribution, biotransformation, cell proliferation, and DNA repair and detoxification between high- and low-dose exposures, between species, and even between tissues. New, ultrasensitive methods of detection make it possible to monitor DNA adducts in animals at exposures below those feasible in chronic bioassays and closer to those expected in the human population. Mathematical models that use such biologic dosimeters might yield more accurate extrapolations and thus improve quantitative risk assessment.

At the present time, few studies of DNA adducts have been carried out under conditions that are appropriate for use in risk assessment. Additional scientific information is needed to improve the extrapolation process, including data from tests using a wide range of doses to determine the saturation points of detoxification and repair, which will represent nonlinearities in the dose-response curves.

The following are some of the subcommittee's conclusions and recommendations regarding the application of DNA-adduct and protein-adduct assays to EPA's assessment of drinking water contaminants:

- DNA-adduct detection methods (especially the radiophosphorus-postlabeling method) have demonstrated the presence of numerous persistent (mostly unknown) DNA lesions in various target cells of untreated animals. The toxicologic importance of those "background" lesions is unknown; they might reflect endogenous exposures to normal body constituents or processes or exposures to naturally occurring mutagens and carcinogens in the environment, such as ultraviolet radiation and mutagens and carcinogens in foods.

- Investigations into the origin and importance of these natural background DNA adducts are needed. The resulting information will be relevant to the interpretation of the impact of DNA alterations induced by occupation, lifestyle, or environmental exposure.

- The presence of DNA adducts or protein adducts might not in itself establish a risk, and tests for their presence should not be used in isolation for hazard or risk assessment.

- If the goal is to assess exposure to a drinking water contaminant identified as genetically toxic, it might sometimes be more appropriate to use protein adducts as dosimeters, although the biologic significance of protein adducts could be quite different from that of DNA adducts. Germ cell studies have suggested that protamine adducts are relevant to genetic risk assessment.

- Baseline data on chemicals in drinking water that are presumed to be genetically toxic should be established, with an eye to revealing qualitative and quantitative associations between DNA adducts or protein adducts and other components of hazard identification.

- Differences between rates of in vivo DNA-adduct formation and repair in somatic and germ cells should be studied. Risk assessment that includes DNA-adduct measurements usually focuses on tumor formation, but other heritable effects of genetic toxicants should be considered a major burden for the human population also.

1

Biologic Significance of DNA Adducts and Protein Adducts

Current evidence suggests associations between the occurrence of adducts formed by specific compounds and various types of toxicity, such as mutation, cancer, and developmental effects. Clinical expression of the toxic effect is usually tissue-specific and can be delayed. DNA adducts form in many tissues, but some of them might be early markers of disease that could be reversed (NRC, 1987). This chapter describes what is known about mechanisms and rates of DNA-adduct formation and removal, the significance of the adduct's position on the DNA, and the correlation of of adducts of certain specific compounds with toxic effects. In addition, protein adducts are discussed as possible markers of exposure.

Studies of laboratory animals and human chemotherapy patients have suggested that DNA adducts can serve as biologic dosimeters in providing estimates of exposure, dose to the target tissue, and sometimes mutagenicity and carcinogenicity (Anderson, 1987; Wogan, 1988). For example, correlations between DNA-adduct formation and exposure, hepatocyte initiation, and hepatocellular carcinoma have been observed in experiments with diethylnitrosamine (Figures 1-1–1-3) (Dyroff et al., 1986), 2-acetylaminofluorene (Beland et al., 1988), aflatoxin B_1 (Croy and Wogan, 1981; Kensler et al., 1986), and N-methyl-4-aminoazobenzene (Tullis et al., 1987). Detection of unique DNA adducts in a population at risk would yield qualitative evidence of exposure. And the use of DNA adducts could perhaps reduce the uncertainty in quantitative risk assessment by providing better dose information for dose-response evaluation. The use of DNA adducts to measure biologically effective dose is scientifically appealing. DNA adducts can indicate a measurable dose at a target site and thus make it possible to bypass—

6

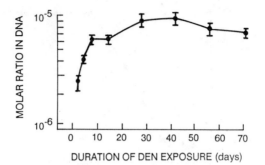

FIGURE 1-1   Relationship of diethylnitrosamine (DEN) exposure to DEN alkylation in 4-week-old Fischer-344 rats. Data points represent mean concentrations of moles of O^4-ethyldeoxythymidine (O^4-EtdT) to moles of deoxythymidine (dT) $\pm$ the standard error of the mean (SEM) for 2–4 animals. Adapted from Dyroff et al., 1986, with permission.

or to confirm—considerations of absorption, distribution, metabolic activation, and detoxification (Hoel et al., 1983).

The estimation of carcinogenic risk usually involves two basic pieces of information (NRC, 1983). A chronic animal bioassay measures the administered doses of a chemical and correlates dose with tumor incidence to provide a quantitative evaluation of carcinogenic hazard at the doses and in

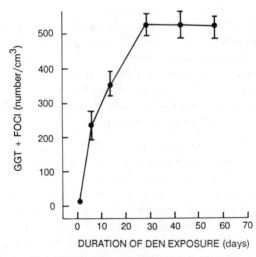

FIGURE 1-2   Relationship of DEN exposure to hepatocyte initiation in 4-week-old Fischer-344 rats. Data points represent mean γ-glutamyl transferase-positive (GGT +) foci per cubic centimeter $\pm$ SEM for 10–12 animals. The plateau in initiation represents a steady state, where the number of newly initiated hepatocytes equals the number of previously initiated hepatocytes that die. Adapted from Dyroff et al., 1986, with permission.

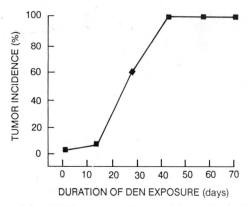

FIGURE 1-3   Relationship of DEN exposure to hepatocellular carcinoma in 4-week-old Fischer-344 rats. Adapted from Dyroff et al., 1986, with permission.

the species tested. Carcinogenic hazard is then combined with information about human exposure to estimate the human risk associated with the chemical. Unfortunately, animal bioassays are limited both practically and economically to measuring tumor incidences at exposures that are much higher than would be acceptable in human populations. Because these models are based on high experimental doses, the resulting data must be extrapolated to permit estimation of the dose-response relationship at doses far below those used in the bioassay. The selection of models that best represent true dose-response relationships in humans at low exposures is controversial. All the mathematical models now used yield similar estimates at high doses, but estimates for low doses deviate widely.

The rates and routes of metabolic activation and detoxification of chemicals differ between sexes, species, and tissues and between high and low doses. Measuring DNA adducts provides one way to understand and even measure those differences. The following are examples:

• Male mice produce different types of DNA adducts from, and more hepatocarcinomas than, female mice after exposure to the same doses of the hepatocarcinogen N-hydroxy-2-acetylaminofluorene (Beland et al., 1982).

• At equimolar doses, rat tissues have higher aflatoxin B_1-adduct concentrations than mouse tissues, possibly because mice have a higher rate of detoxification (Degan and Neumann, 1981; Monroe and Eaton, 1987).

• Rat hepatocytes have a much greater metabolic ability than hepatic sinusoidal cells to activate diethylnitrosamine and thus form DNA adducts (Lewis and Swenberg, 1983).

Dose-dependent changes in rates of metabolic activation and detoxification themselves can affect the relation between administered dose and formation of DNA adducts. For example, the tobacco carcinogen 4-(N-methyl-N-nitrosamino)-1-(3-pyridyl)-1-butanone (NNK) is more efficient per unit dose in producing O^6-methylguanine at low doses than it is at high doses, perhaps because enzymes reach their capacity for activation of a xenobiotic compound. Thus, higher concentrations of the compound do not necessarily result in greater numbers of adducts (Belinsky et al., 1987). In contrast, the efficiency of benzo[a]pyrene (BaP) (Adriaenssens et al., 1983) and formaldehyde (Casanova-Schmitz et al., 1984) in forming DNA adducts and in binding covalently to DNA is greater per unit dose at high exposures, but in a nonlinear fashion.

The effect of DNA repair on DNA-adduct accumulation might also be different at high and low doses. The O^6-alkylguanine DNA alkyltransferases efficiently remove small amounts of the promutagenic adduct O^6-alkyldeoxyguanosine from DNA, but become saturated as the concentration of O^6-alkyldeoxyguanosine in DNA increases (Pegg, 1983). As noted above, species, tissues, and cell types can differ in their concentrations of and abilities to induce these enzymes. For example, human livers have intrinsic concentrations of O^6-alkylguanine DNA alkyltransferase nearly 10 times greater than those in rat livers (Pegg, 1983).

New ultrasensitive methods of detection make it possible to monitor DNA adducts in animals at exposures below those feasible in chronic bioassays and closer to those expected in the human population. Mathematical models that use such biologic dosimeters might yield more accurate extrapolations and thus improve quantitative risk assessment.

Some problems in using DNA adducts to estimate human risks are related to differences between rodents and humans. We can calculate the risk associated with DNA adducts in experimental animals, but interspecies extrapolations remain difficult to validate. Many experiments cannot ethically be performed in humans, and DNA adducts in human target cells or tissues would be expected to vary widely because of individual variations in DNA metabolism and repair.

DYNAMICS OF DNA-ADDUCT FORMATION AND REMOVAL

The chain of causation from toxic chemicals in drinking water or air to alterations of DNA in mammalian cells involves many pharmacokinetic steps. The rate constants of those steps depend on the chemical, species, sex, tissue, and, within a given tissue, cell type. Figure 1-4 shows how metabolic activation and detoxification affect the relationship between external concentration and DNA-adduct concentration in three hypothetical cases of chronic exposure.

The overall estimation of DNA adducts might not be useful, unless one can determine the ratio of biologically important to unimportant adducts.

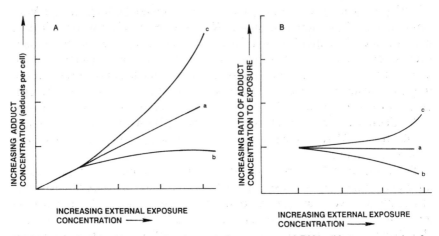

FIGURE 1-4   Relations between chronic external exposures and DNA-adduct concentration for steady state of adduct formation and repair in three hypothetical cases: (a) neither formation nor repair reaches capacity at high concentration; (b) metabolic activation (adduct formation) reaches capacity at high external concentration; (c) DNA repair or detoxification reaches capacity at high concentration. Both scales are linear scales.

The best example of such a classification is demonstrated by the adducts produced by methylating agents; the major DNA adduct formed is N7-methylguanine (N7-MG), but this adduct is not involved in base-pairing and thus is relatively innocuous biologically. A minor adduct, O^6-methylguanine (O^6-MG), which is involved in base-pairing, more closely reflects the mutagenicity and carcinogenicity of methylating agents. The ratio of the two adducts depends critically on the chemical nature of the methylating agent. Hence, the concentration of N7-MG is not particularly useful as a measure of exposure without information on the proportion of N7-MG to O^6-MG and on the nature of the methylating agent itself.

DNA-Adduct Formation Rates

In chronic exposures, the rate of formation of DNA adducts depends on the concentration of compound in the tissue and the rate constant of formation (k_f). The rate of formation varies over time, because of changes in the tissue concentration of reactants that reflect their absorption, transport, and elimination. Low chronic exposures generally do not produce concentrations of xenobiotic compounds at which metabolic activation or detoxification systems reach capacity, so the rate of formation of DNA adducts, dA/dt, can be considered roughly proportional to the concentration of a toxicant that ultimately reacts with DNA, which in turn is proportional to the extracellular concentration of the parent compound. If $\overline{C}$ is the time-weighted average concentration of the toxicant that ultimately reacts with DNA, the average rate of formation of adducts at low chronic exposures is given by:

$$d\overline{A}/dt = k_f\overline{C}, \tag{1}$$

where $\overline{A}$ is the average DNA adduct concentration (e.g., adducts per 10^{10} nucleotides). At high doses, when capacity limitation might be reached, a more elaborate analysis is needed (Travis et al., 1989). The concentration $\overline{C}$ in the target cell might vary with time and tissue. In addition, it will probably vary with the person whose DNA is investigated, because concentrations of activating and detoxifying enzymes vary widely among people.

DNA Repair

DNA adducts are not necessarily stable; some decompose spontaneously at body temperature. For example, alkylation of the nitrogen in purines tends to labilize the glycosidic bond and gives rise to apurinic sites. In addition, enzymatic DNA repair systems can directly remove the adduct itself, remove the DNA base that contains the adduct (base excision repair), or remove nucleotides that contain the adducted base (nucleotide excision repair) (Friedberg, 1985).

The DNA repair systems probably arose as evolutionary consequences of damage to DNA that resulted from ultraviolet radiation (repaired by nucleotide excision), other naturally occurring alkylating agents and mutagens in food (NRC, 1973), and endogenous chemical or enzymatic reactions. The latter reactions are so numerous that, if DNA repair did not occur, 10% of all human DNA bases would be altered in an average lifetime (Tice and Setlow, 1985). The enzymatic DNA repair mechanisms all seem to have capacities far in excess of what is needed to handle the low rate of damage from endogenous reactions and low chronic exposures to most exogenous agents (Table 1-1). At chronic low doses, rates of DNA repair (k_r) are generally limited not by the capacities of repair systems, but by the time for repair enzymes or repair complexes to "find" an adduct. The rate of removal of adducts by repair may be expressed as

$$-d\overline{A}/dt = k_r\overline{A}. \tag{2}$$

For chronic exposures, a steady state is reached when the rate of removal of adducts (Equation 2) equals the rate of production (Equation 1):

$$k_r\overline{A} = k_f\overline{C} \text{ and} \tag{3}$$
$$\overline{A} = (k_f/k_r)\overline{C}.$$

Under conditions of chronic low exposure, the maximal rate of repair is much

TABLE 1-1 Approximate Rates of DNA Damage and Repair in Human Cells at Body Temperature.

Type of Damage	Estimated Occurrences of Damage per Hour per Cell[a]	Estimated Maximal Repair Rate, Base Pairs per Hour per Cell[a]	References
Endogenous			
Depurination	1,000	[b]	Setlow, 1987; Tice and Setlow, 1985
Depyrimidination	55	[b]	Tice and Setlow, 1985
Cytosine deamination	15	[b]	Setlow, 1987; Tice and Setlow, 1985
Single-strand breaks	5,000	2×10^5	Setlow, 1987; Tice and Setlow, 1985
N7-methylguanine	3,500	Not reported	Saul and Ames, 1986
O^6-methylguanine	130	10^4	Setlow, 1987; Tice and Setlow, 1985
Oxidation products	120	10^5	Saul and Ames, 1986; Setlow, 1987
Exogenous			
Background ionizing radiation			
Single-strand breaks	10^{-4}	2×10^5	Setlow, 1987
Oxidation damage	$10^{-4}-10^{-3}$	10^5	Saul and Ames, 1986
Ultraviolet irradiation of skin (noon Texas sunlight)			
Primidine dimers	5×10^4	5×10^4	Setlow, 1987; Tice and Setlow, 1985

[a]Might be higher or lower by a factor of 2 (Setlow, 1983).
[b]Not reported, but the rates are at least 10^4, to judge from the concentration of repair activities in cell extracts.

greater than the rate of introduction of damage (Table 1-1), so the steady-state value of $\overline{A}$ is low. Sensitive techniques are needed to detect these low values.

At low exposure rates, DNA-adduct concentrations are proportional to $\overline{C}$ and hence to exposure concentrations or dose rates. The ratio of $\overline{A}$ to exposure concentration is constant (a curves in Figure 1-4). For exposures at high dose rates, the capacities for adduct formation, detoxification, and repair might be reached. If adduct formation reaches capacity, but repair does not, the rate of formation approaches a constant K_{fmax}; at the steady state,

$$K_{fmax} = k_r\overline{A} \text{ and}$$
$$\overline{A} = K_{fmax}/k_r.$$

$\bar{A}$ is independent of exposure concentration, and the ratio of $\bar{A}$ to exposure concentration approaches zero as the latter continues to increase (b curves in Figure 1-4). However, if detoxification or the repair rate reaches capacity at lower concentrations than the activation rate, adducts continue to increase with time, adduct concentration rises without limit, and the biologic system deteriorates (c curves in Figure 1-4).

DNA that contains adducts has altered template properties, so the rate of introduction of mutations depends on the rate of DNA synthesis. The rate of introduction of altered RNA (possibly leading to changes in gene expression) depends on the rate of transcription. The rates of introduction of errors in replication or transcription depend on both $\bar{A}$ and the rates of replication and transcription. Increased rates of cell replication are frequently associated with high-dose toxicity. Furthermore, DNA synthesis, transcription, and repair vary from one tissue to another and from one subject to another. The magnitudes of the variations depend on the particular repair system involved, genetic and environmental factors, and the pharmacokinetic and toxic properties of the chemical agent producing the adducts (Wogan, 1988).

In bacterial systems, exposure to mutagens at low concentrations often induces synthesis of new repair enzymes and an increase in repair rate. Such an adaptation is well documented for ultraviolet irradiation, whose effects are repaired by nucleotide excision (Friedberg, 1985, pp. 431, 432). An increase in the rate of repair of DNA damage can also be produced by alkylating agents and such other agents as benzo[a]pyrene that yield high-molecular-weight (bulky) DNA adducts. Adaptation increases the value of k_2 in Equation (2) and results in a decrease in the steady-state value of $\bar{A}$. Adaptation reactions in human cells have not been well documented.

Insofar as some DNA adducts have been shown to be important in mutagenesis and carcinogenesis, estimates of long-term risk would be expected to be proportional to the steady-state concentration of such adducts. The constant of proportionality depends not only on rates of transcription of RNA and replication of DNA, but on biologic factors, such as the location of adducts in the genome and the presence of endogenous promoters or inhibitors.

SITE RELEVANCE

Many carcinogens and mutagens react at more than one site on DNA, producing several types of DNA adducts (Figure 1-5). As stated above, adducts at different sites can differ greatly in the rates at which they are formed and repaired and in their efficiency in causing mutations. Thus, data on overall covalent binding or a covalent binding index (Lutz, 1979) could be misleading. It is important to consider all available relevant biologic data,

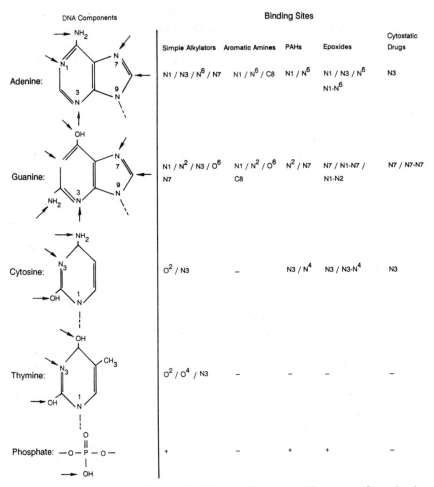

FIGURE 1-5 Potential sites of binding in DNA. Specific nitrogen (N), oxygen (O), and carbon (C) atoms on the DNA components have different susceptibilities to binding. Adapted from Singer (1985) with additional information from Beland and Kadlubar (1985), Delclos et al. (1987), and Hemminki (1983).

including mutagenic efficiency, when choosing DNA adducts to be used as molecular dosimeters or for risk assessment.

Alkylation

In DNA, the N7 position of guanine is the most nucleophilic site, and it is by far the site most often alkylated by electrophiles. All the ring nitrogens of the DNA bases, except the nitrogen attached to the deoxyribose sugar, have been shown to be alkylated to some extent by a variety of agents (Singer, 1975). Figure 1-5 shows all the potential sites for alkylation in the four bases found in DNA, as well as on its phosphate backbone. These sites include the N1, N3, N7, and C8 of guanine; the N1, N3, N7, and C8 of adenine; the N3 of thymine; and the N3 of cytosine. In addition, all the exocyclic nitrogens and oxygens can be alkylated; these sites include the N^2 and O^6 of guanine, the N^6 of adenine, the O^2 and O^4 of thymine, and the O^2 and N^4 of cytosine. Some chemicals, such as ethyl nitrosourea (ENU), have also been shown to alkylate the phosphate oxygens on the DNA backbone, forming phosphotriesters. With ENU, about 60% of total DNA ethylation occurs on the phosphate group (Singer, 1982).

All the nucleophilic sites in DNA mentioned above are potential sites of alkylation, as determined by in vitro experiments, but not all are significantly affected in vivo. Configuration and secondary structure of DNA can play a major role in chemical reactivity (Brown, 1974; Singer and Fraenkel-Conrat, 1969). Other factors, such as the size of the binding electrophile and the association of proteins with chromosomal DNA, also appear to affect the sites or magnitude of DNA alkylation in vivo (Singer, 1982; Swenson and Lawley, 1978).

Although many chemicals can alkylate DNA directly, others, such as aromatic amines and polycyclic aromatic hydrocarbons, often undergo complex enzymatic modifications before they can alkylate DNA (Brookes, 1977; Kriek and Westra, 1979; Miller, 1978; Sims and Grover, 1974). There are some striking differences between the DNA adducts produced by enzymatically modified chemicals and the adducts formed by simple alkylating agents (Hemminki, 1983). Not only are many of the adducts formed by enzymatically modified chemicals large and aromatic, but for polycyclic aromatic hydrocarbons, the preferred site of reaction in DNA is different. They generally alkylate exocyclic amino groups, particularly the N^2 of guanine, whereas the preferred site of aromatic amines is the C8 of guanine.

Base Mispairing

During DNA replication and in newly synthesized DNA, hydrogen bonds become less stable, and mispairing can occur; thus, alkylation of the DNA bases at sites involved in hydrogen binding is potentially mutagenic (Kroger and Singer, 1979; Singer et al., 1978a, 1979, 1983a,b). Those sites include the N1, N^2, and O^6 of guanine; the O^2, N3, and N^4 of cytosine; the N1 and N^6 of adenine; and the N3 and O^4 of thymine. For example, alkylation of the O^6 of guanine can cause miscoding by DNA and RNA polymerases (Abbott and Saffhill, 1979; Gerchman and Ludlum, 1973). O^6-Alkylguanine has been shown to direct the misincorporation of substantial amounts of thymine, instead of the expected cytosine, into newly synthesized DNA (Abbott and Saffhill, 1979; Green et al., 1984; Lawley, 1974; Loechler et al., 1984). There is also evidence that O^6-alkylguanine can direct some misincorporation of adenine (Snow et al., 1983). Bulky adducts distort the DNA, again increasing the likelihood of misincorporation.

Hydrolysis

The N3 and N7 alkylpurines can be hydrolyzed from DNA as a consequence of the instability of their glycosyl bonds, even at neutral pH. The half-lives of those adducts in DNA can range from a few hours to several days (Singer and Grunberger, 1983). Their rates of spontaneous hydrolysis are about 10^6 times greater than the rates for the unmodified purines. The glycosyl bonds of pyrimidines are 100 times more stable than those of the purines. As a consequence, depyrimidination of even the most labile alkylpyrimidine, O^2-alkylcytosine, has a half-life about 35 times that of N7-alkylguanine (Singer et al., 1978b). Nevertheless, depyrimidination of O^2-alkylcytosine can contribute significantly to the formation of apyrimidinic sites. If apurinic or apyrimidinic sites are present in DNA at the time of replication, any base can be misincorporated into the newly synthesized DNA opposite the gap in the parental strand (Lawley and Brookes, 1963).

Phosphate Adducts

The formation of alkyl phosphotriesters, first measured by Bannon and Verly (1972) and later by Sun and Singer (1975), on the phosphate backbone of DNA does not make the chain unstable. Alkyl phosphotriesters have been reported to repair with a half-life of several days in rat liver (O'Connor et al., 1973, 1975) and rat brain (Goth and Rajewsky, 1974), perhaps as a result of enzymic excision of these products. Miller et al. (1971, 1974) and Kan et al. (1973) reported that triesters exhibit changes in a number of properties that are likely to affect normal replication. However, Rajewsky et al. (1977) found no correlation between the persistence of phosphotriesters in DNA of brain and liver and the sensitivity of these organs to carcinogenesis by ENU.

Cross-Links

DNA–DNA cross-links can be created by bifunctional or polyfunctional alkylating agents. Brookes and Lawley (1961) demonstrated that di(guanin-7-yl) derivatives could be formed in DNA exposed to bifunctional alkylating agents. The cross-linking is normally expected to occur between guanines on opposite strands of DNA. Formation of such an adduct is generally believed to be lethal, since it would prevent DNA strand separation at replication. However, no evidence has been presented to relate the occurrence of mutations or cancers to the formation of dialkyl-base adducts.

DNA ADDUCTS AND TOXIC EFFECTS

Relating information concerning DNA-adduct site and molecular biologic consequences of adduct formation to multistep processes like mutagenesis and carcinogenesis is difficult at best. Specific toxic effects of specific DNA adducts must be correlated with the induction of gene mutation, germ cell mutation, or tumor formation in animal models before the impact of DNA-adduct formation can be assessed in humans. DNA-adduct dosimetry studies are available for different chemical classes of mutagens and carcinogens in various bioassays, including in vitro short-term and single- and multiple-dose whole animal exposures. Some specific carcinogen–DNA-adduct relationships have been demonstrated in humans exposed to carcinogens occupationally, environmentally, or otherwise.

In Vitro Short-Term Bioassay

The correlations between cytotoxicity, mutation frequency, and binding of BPDE I, which is the anti-isomer of benzo[a]pyrene (BaP) diol epoxide (BPDE), to DNA have been examined in normal diploid human fibroblasts in culture and xeroderma pigmentosum cells (Yang et al., 1980; McCormick and Maher, 1985), the latter of which are deficient in excision repair. Those studies seem to show that BPDE I-deoxyguanosine (BPDE I-dG) caused the observed cytotoxicity and mutations and that the relationship between BPDE I-DNA binding and frequency of induced mutations (in tests for resistance to the toxic effects of 6-thioguanine) is linear in normal cells. Similar linear relationships between mutation frequency and binding of BPDE I or BPDE II (the isomer of BPDE) to DNA have been reported in *Salmonella typhimurium* strains TA98 and TA100 (Fahl et al., 1981). Newbold et al. (1979) used an in vitro short-term bioassay to construct curves of mutation frequency versus DNA binding for 7-bromomethylbenz[a]anthracene and BPDE I. They demonstrated linear to curvilinear relationships between total DNA binding in Chinese hamster V79 cells and mutation frequency (in tests for resistance to the toxic effects of 8-azaguanine) in the same cell type.

A study by Arce et al. (1987) related overall BaP–DNA-adduct and BPDE I-dG–DNA-adduct concentrations to a variety of end points in four different cellculture systems: gene mutation and sister-chromatid exchange in Chinese hamster V79 cells, gene mutation and chromosomal aberrations in mouse lymphoma L5178Y cells (thymidine kinase [TK + / −]), virus transformation in Syrian hamster embryo cells, and structural transformation in mouse embryo C3H10T$_{1/2}$ fibroblasts. The relationship between the genetically toxic effect and BaP–DNA binding or BPDE I-dG was linear in each assay. Each genetic end point was induced with a different efficiency on a per-adduct basis, whether expressed as total BaP–DNA binding or as specific BPDE I-dG–DNA adducts. Even when standardized by expression of the number of BPDE I-dG adducts per unit of DNA required to double the frequency of induced biologic response, results were the same; the BPDE I-dG–DNA adduct had different potencies in different cell cultures and for different biologic end points. Similar results have been obtained with aromatic amines and amides (Arce et al., 1987; Heflich et al., 1986).

Studies of the relationships between carcinogen-DNA adducts and mutagenesis have not been limited to high-molecular-weight (bulky) aromatics like BaP. Van Zeeland et al. (1985) reported on the relationships between O^6-ethylguanine (O^6-EG) formation induced by ethylating agents in Chinese hamster V79 cells (in tests for resistance to the toxic effects of 6-thioguanine), in *Escherichia coli* (in tests for nalidixic acid resistance), and in mice (specific-locus mutations). For each compound, very similar mutation frequencies per locus were observed in all three assays; this suggests that O^6-EG might be a good marker for monitoring exposure to chemical mutagens and for predicting mutation induction by a methylating agent.

In Vivo Germ Cell Mutation

Little is known about the effect of DNA adducts on germ cell mutation. In two studies of male germ cells, the frequency of sex-linked recessive lethal mutations in *Drosophila melanogaster* (Aaron and Lee, 1978) and of specific-locus mutations in mice (Van Zeeland et al., 1985) increased linearly with increasing DNA ethylation. Studies comparing the effects of single and repeated exposures were not as clear-cut. Russell (1984) and Russell et al. (1982) found that ENU was a potent inducer of specific-locus mutations in mouse spermatogonial stem cells and that the mutation frequency was 5.8 times greater after a single 100-mg/kg exposure than after 10 weekly 10-mg/kg exposures.

In contrast, molecular-dosimetry studies with ENU (Sega et al., 1986) found that the amount of O^6-EG formed in testicular DNA by ENU at 100 mg/kg was only 1.4 times that expected in response to 10 weekly 10-mg/kg exposures. That result did not support the idea that O^6-EG was an important mutagenic lesion in the germ cells. In fact, the molecular-dosimetry data fit the genetic data well, if two hits on the DNA, not involving the O^6 of guanine, are assumed necessary to produce an effect. Thus, in correlating DNA-adduct formation with in vivo germ cell mutation, one must consider both acute and chronic exposure. This has been clearly demonstrated with regard to tumorigenesis as the end point.

Considerable evidence is accumulating that, for many agents, the induction of mutations does not occur randomly over the chromosome, but that a variety of mutations are formed at preferential "hot spots" in the chromosome (e.g., Benzer, 1961; Drobetsky et al., 1987; Richardson et al., 1987; Skopek et al., 1985; Thilly, 1985; Vrieling et al., 1988). Whether they differ between high and low doses and between species is not yet known, but the effect may be due to site-specific variations in DNA-adduct formation and repair.

Tumorigenesis in Animal Models

Models of tumor initiation, promotion, and progression have been developed from studies in experimental carcinogenesis. In those models, DNA adducts appear to be of prime importance in the initiation stage of tumor formation, although DNA lesions might also facilitate conversion of initiated cells into tumors or of benign cells into malignant cells.

Tumor initiation is often considered to be a single first step in tumor formation. However, initiation is a complex, multistep process. The electrophilic reactivity of chemicals or their metabolites with DNA does not lead stoichiometrically to mutation or cancer. Cell replication must occur prior to repair of DNA adducts for mutations to be induced. Many types of DNA lesions are induced by genetically toxic agents, and all lesions are not repaired equally. Information about these relationships is generally scanty, except for the case of some monofunctional alkylating agents, where the induction of DNA damage and the ultimate induction of mutation in target cells are directly related quantitatively. Persistent DNA lesions should be considered good markers for measuring exposure; whether or not they are also markers of tumor initiation depends on the properties of the specific genetic toxicant of concern.

Many carcinogenic compounds form DNA adducts that can be measured in rodent tissues after single and multiple doses or after chronic exposure (Bedell et al., 1982; Boucheron et al., 1987; Poirier and Beland, 1987; Swenberg et al., 1984; Wogan and Gorelick, 1985). Detecting all DNA adducts derived from a single carcinogen is complex, however, because they can form at many sites on all four DNA bases and on the phosphate backbone of DNA, and they are repaired at different rates (Wogan, 1988). Few studies have shown a correlation between adduct formation early in carcinogen exposure and tumor formation in the same biologic system (Beland et al., 1988; Croy and Wogan, 1981; Dyroff et al., 1986; Kensler et al., 1986; Tullis et al., 1987), but maximal total adduct levels in target tissues of laboratory animals usually reflect carcinogen potency and may or may not be linearly related to dose over a wide range (Swenberg et al., 1987; Wogan, 1988).

This suggests that DNA adducts are generally necessary but not sufficient for tumor formation (Beland et al., 1988; Branstetter, 1987; Neumann, 1983;

Wogan and Gorelick, 1985). Thus, some adducts might be found in organs that are not targets for tumorigenesis, and the same or smaller numbers of adducts might be found in organs that are targets (Beland et al., 1988). For example, Goth and Rajewsky (1974) and Kleihues and Margison (1974) demonstrated that O^6-alkylguanine persisted in rat brain, but not in liver or kidney, after exposure to ENU or methylnitrosourea (MNU), which primarily produce brain tumors. Kleihues et al. (1974) used methyl methanesulfonate, a chemical much weaker than MNU in inducing brain tumors, to demonstrate that tumor induction was proportional to the amount of O^6-methylguanine (O^6-MG) in the brain. Other experiments implicating O^6-alkylguanine as a potentially mutagenic and carcinogenic lesion have been reported (Bedell et al., 1982; Beranek et al., 1983; Cairns et al., 1981; Dodson et al., 1982; Frei et al., 1978; Lawley and Martin, 1975; Lewis and Swenberg, 1980; Newbold et al., 1980; Swenberg et al., 1982). However, Kleihues and Rajewsky (1984) showed that persistence of O^6-MG does not always result in the production of brain tumors. Mouse, gerbil, and hamster brains can show concentrations of O^6-MG similar to those in rats, but have low susceptibility to brain-tumor induction by MNU.

Even with today's supersensitive analytical methods, DNA adducts are sometimes unmeasurable at doses that are tumorigenic (Boucheron et al., 1987). Furthermore, tumors can occur spontaneously without chemical exposure and known DNA-adduct formation. In addition, the persistence of DNA adducts may or may not be related to susceptibility and target tissue specificity (Wogan, 1988). Thus, the relationship between DNA-adduct formation and tumorigenesis is by no means clearly established.

Differences in DNA-adduct formation and persistence appear to provide an explanation for some target site specificities in carcinogenesis (Beland and Kadlubar, 1985; Swenberg and Fennell, 1987). For example, rats are more susceptible than mice to the carcinogenic effect of aflatoxin B_1, and the difference is correlated with the relative concentrations of aflatoxin–DNA adducts in the liver (Croy and Wogan, 1981). Cell specificity of tumor induction in both mice (liver) and rats (liver and lung) has been attributed to differences in DNA-adduct accumulation and persistence among various cell types (Belinsky et al., 1987; Lewis and Swenberg, 1980; Lindamood et al., 1982).

Tumors induced by chronic exposures to methylating hepatocarcinogens are predominantly hemangiosarcomas (involving nonparenchymal cells), whereas exposures to ethylating agents cause hepatocellular carcinomas (involving hepatocytes) (Richardson et al., 1985; Swenberg et al., 1984). In the case of methylating agents, the nonparenchymal cells accumulate O^6-MG, and the hepatocytes do not. Exposure to ethylating agents, however, leads to accumulation of O^4-ethylthymine (O^4-ET), but not O^6-ethylguanine (O^6-EG), in hepatocytes. Thus, it appears that O^4-ET, but not O^6-EG, is correlated with occurrence of cancer in hepatocytes of rats. This primarily reflects differences in the abilities of these cell types to repair O^6-alkyl-guanine.

In vivo studies of BaP binding and tumorigenesis performed on mouse skin (Ashurst et al., 1983; Cohen et al., 1979) and other organs (Adriaenssens et al., 1983; Anderson et al., 1981; Stowers and Anderson, 1985; Wogan and Gorelick, 1985) yielded mixed results. BaP–DNA binding was observed in tissues both susceptible and nonsusceptible to the tumorigenic effects of BaP. Such results indicate that the presence of adducts alone might not necessarily lead to tumor formation.

Potential germ cell mutagenesis must be regarded as a major burden for the human population (NRC, 1986, p. 69), even though heritable mutations caused by chemical exposures have not been detected in humans. In somatic cells, cancer is a heritable process, and DNA carries heritable information. DNA adducts might be sensitive indicators of early genetic effects that could be correlated with chromosomal damage in the target organ. Data from in vivo experiments on mouse liver with benzidine–DNA adducts (Talaska et al., 1987) support the hypothesis that carcinogen–DNA adducts induce chromosomal aberrations and perhaps other toxic effects, including neoplasia.

Single-Dose Exposures of Animals

In contrast with the uncertainty of the relationship between DNA-adduct formation and tumorigenesis, the presence of unique, measurable DNA adducts always indicates that exposure has occurred. In animals, single doses of carcinogens that do not saturate metabolic activation, detoxification, or DNA repair are linearly correlated with the numbers of DNA adducts measured in different organs (Figure 1-4A, curve a). That is true of a wide range of doses of BaP given either topically (Pereira et al., 1979; Shugart, 1985) or orally (Dunn, 1983). Similar studies with aflatoxins administered orally (Wild et al., 1986) or intraperitoneally (Appleton et al., 1982) have shown liver DNA adducts to increase stoichiometrically with dose. When metabolic activation reaches capacity, a less than linear dose response is evident (Appleton et al., 1982); and when detoxification or DNA repair reaches capacity, a greater than linear dose response occurs (Adriaenssens et al., 1983; Casanova-Schmitz et al., 1984).

Chronic Exposures of Animals

Dose-response curves for DNA adducts formed in animals during chronic carcinogen administration have characteristic profiles that reflect the combined processes of DNA-adduct formation and their removal by DNA repair systems. Several studies have demonstrated that exposure by continuous feeding or frequent injection generally results in a DNA-adduct dose-response curve that is initially linear but levels off at higher doses, where the net rate of change in the concentration of adducts reaches a steady state (Figure 1-4A, curve b). That is true for the liver carcinogens aflatoxin (Croy and Wogan, 1981; Wild et al., 1986), 2-acetylaminofluorene (Poirier et al., 1984), and diethylnitrosamine (Boucheron et al., 1987) and for the lung, nasal mucosa, and liver carcinogen 4-(*N*-methyl-*N*-nitrosamino)-1-(3-pyridyl)-1-butanone (NNK) (Belinsky et al., 1986, 1987). In contrast, when Neumann (1984) injected *trans*-4-acetylaminostilbene into rats at 3- to 4-day intervals, he found that DNA-adduct formation in four organs had not reached a plateau within 6 weeks. It has been suggested that the failure to reach a plateau might be due to the time between exposures, as well as to the liver's relative efficiency in removing adducts formed from aromatic amines (Poirier et al., 1984).

Chronic feeding studies show that the plateau might vary with the concentration of carcinogen in the diet. For example, when diethylnitrosamine was given continuously to male Fischer-344 rats in drinking water, an increase in concentration from 0.4 ppm to 40 ppm caused a 100-fold increase in the steady-state adduct concentration in the livers of male Fischer-344 rats; however, further exposures up to 100 ppm did not cause any additional increase in adduct formation (Boucheron et al., 1987). This nonlinearity in dose-response primarily reflects killing of cells that contain adducts. Exposure to 100 ppm of diethylnitrosamine causes a 20-fold increase in liver cell proliferation. Again, the presence of DNA adducts indicates carcinogen exposure; but the magnitude of exposure cannot be determined from a single adduct measurement, if the sample is taken after either formation or removal of adducts has exceeded the straight linear portion of the dose-response relationship in either dose or time.

Studies in Humans

In many studies in which DNA adducts have been measured in humans, the exposure was to a ubiquitous compound producing delayed clinical effects, thus making cumulative exposure unknown and unexposed control populations unidentifiable. However, cancer patients receiving chemotherapy are one human cohort in which drug dosage is known and unexposed controls are easy to identify. Cisplatin is a known carcinogen in rodents (Leopold et al., 1979), but has not been reported to produce secondary tumors in cancer patients receiving platinum-based chemotherapeutic agents. Nucleated blood cell adducts in DNA of testicular- and ovarian-cancer patients receiving cisplatin have been measured, and a dose-response relationship for adduct formation has been observed (Poirier et al., 1985, 1987). Although only about 60% of the persons studied had measurable adducts, the adducts were correlated with dose; but the absence of adducts in other patients was not related to absence of exposure. In fact, the presence of adducts was shown to be correlated with tumor remission (Reed et al., 1987)—a finding of clinical importance.

A limited number of carcinogen–DNA-adduct relationships have been successfully demonstrated in humans with occupational or environmental carcinogens; however, in many of these studies, information about dose was scanty or nonexistent, and the results obtained had to be interpreted carefully because of the nature of the assay system used. For example, an antiserum against BaP–DNA adducts has substantial cross-reactivity with other PAH–DNA adducts. Therefore, whenever this antiserum is used with an enzyme-linked immunosorbent assay (ELISA) or an ultrasensitive enzymatic radioim-munoassay (USERIA) to detect BaP–DNA adducts, it is actually detecting a variety of PAH–DNA adducts. This antiserum was used in a pilot study (Perera et al., 1982) in which lung tissues of patients with and without lung cancer were assayed by ELISA. The results showed that lung DNA from 4 of 14 patients with lung cancer was positive in this test, indicating the presence of adducts, probably from a variety of hydrocarbons. Shamsuddin et al. (1985) used the same antiserum in a USERIA to measure BaP–DNA antigenicity in the white blood cells of roofers and foundry workers, two groups known to have substantial occupational exposure to BaP. Samples from 7 of 28 roofers and from 7 of 20 foundry workers were positive for BaP–DNA antigenicity. Samples from 2 of 9 volunteer controls were also positive, and both of those were smokers. In a cross-sectional analysis that lacked controls, Harris et al. (1985) detected BaP–DNA antigenicity in the peripheral lymphocytes of coke-oven workers using a USERIA and syn-chronous fluorescence spectrophotometry (SFS) with the same antiserum. Of 27 coke-oven workers, 18 showed BaP–DNA antigenicity with a USERIA and 9 did not; of 41, 31 showed detectable adduct formation with SFS and 10 did not. Antibodies to BaP–DNA adducts were found in the serum of 28% of the workers (Harris et al., 1985).

Haugen et al. (1986) studied BaP–DNA antigenicity in peripheral lym-phocytes of 38 top-side coke-oven workers and conducted personal air sam-pling for PAHs inside and outside the respirators of 4 of those workers. By SFS, samples from 4 of 38 workers had putative adducts; and by USERIA, 13 of 38 samples had detectable antigenicity. Inside the respirators, the concentration of total PAH ranged from 51 to 162 $\mu g/m^3$, and the concen-tration of BaP ranged from 1 to 4 $\mu g/m^3$.

Monoclonal antibodies have been used to detect O^6-methyldeoxyguanosine adducts (presumably produced from nitrosamines formed in food) in the DNA of esophageal and stomach tissue of subjects from China (Umbenhauer et al., 1985). A total of 37 malignant and nonmalignant specimens were ob-tained from esophageal and stomach tissue of patients undergoing surgery for cancer of the esophagus in Linxian Province, China (an area of high incidence of esophageal and gastric cancer), for study with a radioimmunoassay to detect the presence of O^6-methyldeoxyguanosine adducts. Additionally, 12 human tissue samples from Europeans were studied. Of the 37 samples from China, 27 had detectable adducts, and 5 of the 12 European samples had detectable adducts.

Perera et al. (1987a) used ELISA to measure BaP–DNA antigenicity in multiple samples of white blood cells of 22 smokers and 24 nonsmokers. They also assayed 4-aminobiphenyl-hemoglobin adducts with negative chemical-ionization mass spectrometry. In sample 1, 5 of 22 smokers (23%) and 7 of 24 nonsmokers (29%) showed detectable BaP–DNA antigenicity; in sample 2, 4 of 20 smokers (20%) exhibited detectable antigenicity; in sample 3, 4 of 21 smokers (19%) and 4 of 21 nonsmokers showed detectable antigenicity. Of the subjects with PAH–DNA adducts, the quantity of adducts was higher in smokers than in nonsmokers and higher in women than in men. Data on exposure of nonsmokers to environmental tobacco smoke were not available. 4-Aminobiphenylhemoglobin adduct formation correlated best with indexes of active smoking.

More recently, Perera et al. (1987b) have utilized an ELISA to study BaP–DNA antigenicity among 22 foundry workers and 10 non-occupationally exposed controls. Foundry workers were characterized as having high, medium, and low exposures. Mean levels of PAH–DNA adducts (femtomoles/microgram) increased with exposure (low = 0.32, medium = 0.53, high = 1.2), and there was a significant difference between control (0.06) and pooled exposure (0.60) group means.

^{32}P-postlabeling has been used to investigate the presence of DNA adducts in placentas from smokers and nonsmokers (Everson et al., 1986). Several modified nucleotides were detected with ^{32}P-postlabeling; one was strongly related to maternal smoking, but only weakly related to either historical or biochemical measures of intensity of smoking.

^{32}P-postlabeling has also been used to evaluate formation of PAH–DNA adducts among nonsmoking pregnant women with exposure to residential wood combustion smoke (RWC), and among unexposed, nonsmoking pregnant women (Reddy et al, 1987). DNA was isolated from specimens from 12 exposed women (8 white blood cells, 4 placentas) and specimens from 13 unexposed women (8 white blood cells, 5 placentas). Comparison of exposed subjects with controls did not reveal exposure-related adducts; however, all placentas contained unidentified adducts which were not present in maternal white blood cells.

Phillips et al. (1988) studied the same group of foundry workers examined by Perera et al. (1987b), utilizing ^{32}P-postlabeling. Foundry workers were classified as having high (>0.2 μg BP/m^3), medium (0.05–0.2 μg BP/m^3), or low (<0.05 μg BP/m^3) BaP exposure based on historical industrial hygiene measurements and job title. Aromatic adducts were found in DNA from 3 of 4 samples from the high-exposure group, 8 of 10 samples of the medium-exposure group, 4 of 18 samples from the low-exposure group, and 1 of 9 samples from unexposed controls. No differences due to smoking habits were observed.

Genetic heterogeneity of metabolic activation and exposures to exogenous compounds that may alter metabolism and adduct formation will contribute to interindividual variability in adduct formation. For example, heterogeneity of human polycyclic aromatic hydrocarbon activation is well established, and the extent of DNA-adduct formation varies over a 1,000-fold range (Harris et al., 1982). In addition, some compounds, such as antioxidants and other dietary ingredients, modulate metabolism and thereby alter adduct formation. Ethoxyquin, an antioxidant in cabbage and other plants, decreased adduct formation in rats by 95% when ingested with aflatoxin B_1 in a rat-liver tumorigenesis study (Kensler et al., 1985). A concomitant decrease in preneoplastic foci was observed. It was concluded that the cabbage content of the diet may be important in the relationship between aflatoxin–DNA adducts and liver cancer in the Chinese.

PROTEIN ADDUCTS

In protein, the amino acids most likely to be alkylated are cysteine, histidine, lysine, and the N-terminal amino acid. Hemoglobin adducts were suggested as suitable for monitoring dose by Osterman-Golkar et al. in 1976; unlike DNA adducts, which can be removed by repair mechanisms, protein adducts had been observed to persist over the 40-day lifespan of the erythrocyte in the mouse (Osterman-Golkar et al., 1976). Ehrenberg and Osterman-Golkar (1980) reviewed the use of protein alkylation to detect mutagenic agents. Such use requires that the exposure result in stable, covalent derivatives of amino acids; that the target protein be found in easily accessible fluids, such as blood; and that the derivatives be present in adequate concentrations.

Pereira and Chang (1981), studied the ability of 15 carcinogens and mutagens in a wide range of chemical classes to bind covalently to hemoglobin in rats. They used radiolabeled test compounds to demonstrate covalent binding of all the mutagens and carcinogens to hemoglobin. The extent of binding with the different compounds had a range of a factor of 100, but the fact that protein adducts were formed from all the compounds studied indicates the potential usefulness of this molecular target for measuring exposure.

Segerbäck et al. (1978) showed that alkylation of hemoglobin in mice by methyl methanesulfonate (MMS) was a linear function of the injected dose. Similarly, the production of N-3-(2-hydroxyethyl)histidine in hemoglobin was found to be a linear function of ethylene oxide (EtO) inhalation exposure (Osterman-Golkar et al., 1983). Dose-response relationships have also been established for 4-aminobiphenyl (Tannenbaum et al., 1983), which showed a linear rate of binding to hemoglobin over a 10,000-fold exposure range; for trans-4-dimethylaminostilbene (Neumann et al., 1980), which had a linear rate of binding to rat hemoglobin over a 100,000-fold exposure range; and for chloroform (Pereira and Chang, 1982), which showed a linear rate of binding to rat and mouse hemoglobin over a 1,000-fold exposure range.

Although the amount of hemoglobin alkylation can be related to chemical exposure, it can be used as an indication of risk of genetic toxicity only if hemoglobin alkylation is correlated with alkylations at mutationally important targets, such as DNA. For example, Neumann et al. (1980) showed that the binding of trans-4-dimethylaminostilbene to plasma proteins and hemoglobin was proportional to its binding to liver DNA, and Pereira et al. (1981) found that dose-response curves for the binding of 2-acetylaminofluorene to rat hemoglobin and to liver DNA were closely related over a large dose range. Thus, it might be possible to estimate DNA binding through measurement of protein binding.

A more direct use of protein adducts to estimate the risk of genetic toxicity is to select a protein that can be a genetically significant target. Sega and Owens (1978, 1983) found that exposure of male mice to ethyl methanesulfonate (EMS) and MMS produced significant increases in alkylation in late spermatids and early spermatozoa (the most genetically sensitive germ cell stages) that could not be attributed to increased DNA alkylation but were correlated with sperm protamine alkylation and dominant lethal mutations.

Sega and Owens (1987) found that the temporal pattern of alkylation produced by ethylene oxide in protamine, but not DNA, of the maturing sperm stages of mice is correlated with the pattern of dominant lethal mutations produced by ethylene oxide in the same stages. Thus, measurement of chemical adducts in human sperm protamine might be a useful means of assessing human germinal exposure to genetic toxicants.

Sega (Biology Division, Oak Ridge National Laboratory, Oak Ridge, Tenn., personal communication, 1987) also studied the binding of ^{14}C-acrylamide to developing spermiogenic stages in mice. The temporal pattern of acrylamide binding in the different spermiogenic stages paralleled the temporal pattern of induced dominant lethal mutations and heritable translocations noted by Shelby et al. (1986, 1987), with the greatest binding in late spermatid and early spermatozoal stages. Binding of acrylamide to DNA was not statistically measurable in the different stages, and binding to protamine could account for essentially all the germ cell alkylation.

The above-described studies of methyl methanesulfonate, ethyl methanesulfonate, ethylene oxide, and acrylamide provide compelling evidence that protamine alkylation is temporally associated with dominant lethal mutations.

Sega noted that the proportion of DNA adducts formed in the sensitive stages of spermiogenesis is small (e.g., MMS, EMS, and EtO) or not measurable (e.g., acrylamide). However, late spermatogenic cells are known to be repair-deficient, and it is possible that dominant lethal mutations occur because a small number of DNA lesions remain unrepaired. Further research is needed to investigate the mechanism by which these low-molecular-weight mutagenic compounds cause dominant lethal mutations and elucidate the relative roles of protamine alkylation and DNA alkylation.

SUMMARY

To use DNA adducts in risk estimation, one must relate them to other biologic events, such as germ cell mutation, tumorigenesis, or developmental effects. Experimental data correlating tumorigenesis with profiles of DNA-adduct dosimetry in the same animal tissues are sparse (they include studies on diethylnitrosamine and the liver carcinogens 4-(N-methyl-N-nitrosamino)-1-(3-pyridyl)-1-butanone, 2-acetylaminofluorene, and aflatoxin). Some correlations have been observed between persistence of DNA adducts in target tissues and the induction of tumors, but with some compounds no correlations have been noted. This probably reflects the need to incorporate more biologic processes than DNA-adduct formation into risk assessment. No proof exists that developmental effects occur in humans; however, they are presumed to represent a percentage of the genetic damage known to occur.

One immediate problem is the lack of appropriate data sets from which models can be constructed and validated. Both acute and chronic testing should be performed over a wide dose range to acquire knowledge of the points at which detoxification and DNA repair reach their capacities and thus cause nonlinearities in dose-response relationship curves. Dose-response relationships for single exposures over a dose range of 10^3 have been established for tumor induction on only three carcinogens: dimethylnitrosamine, diethylnitrosamine, and benzo[a]pyrene. Several compounds have been studied in bioassays in which the dose ranged over a factor of 100, but bioassays on most carcinogens use doses that range over a factor of 10 or less— including the largest study ever performed, the effective-dose (ED_{01}) bioassay of 2-acetylaminofluorene (Staffa and Mehlman, 1979). Few DNA-adduct studies have covered dose ranges and used exposure protocols that could be compared. Although a broader dose range is not always possible because of the occurrence of toxic effects, present adduct-detection methods are probably now capable of measuring the results of testing with very low doses.

Correlations between DNA-adduct dose-response relationships and biologic effects seem to be compound-specific and independent of chemical class or biologic end point. Even for a single compound, quantitative comparisons of chemical-DNA binding and hazard assessment are complicated. One relationship will not accurately describe all situations; it will vary with the compound, the specific target tissue, the organism's exposure history, the duration and time of exposure, etc. Individual rates of metabolic activation of carcinogens (particularly PAHs) and repair capacities are variable and moderated by personal exposure histories. Thus, because the same chemical exposure can produce widely varying numbers of adducts, prediction of the extent of exposure or the resultant cancer risk is much more difficult in humans on the basis of DNA adducts than in homogeneous laboratory animals. In addition, for many toxic chemicals, the mutagenic or tumorigenic adduct has not been identified and can occur among many others that may not produce deleterious effects; thus, measuring overall DNA binding attributable to a specific chemical could lead to errors in the estimation of hazard.

Despite current gaps in knowledge, DNA-adduct research represents a very promising means to improve risk assessment. When more extensive data become available, they might be used in individual risk assessment to confirm suspected exposures, improve estimates of target tissue dose, and reveal metabolic activation and detoxification parameters that moderate the formation of DNA adducts by a specific carcinogen. In general risk assessment, they could be valuable in estimating dosimetry and systemic distribution and in establishing possible target tissues or organs and the potential for irreversible toxicity, such as cancer, mutation, or developmental effects. They might improve estimates of the rates of tumor and adduct formation in animals in response to low doses on the basis of high-dose effects and provide better models for predicting mechanisms in humans. Large-scale DNA-adduct dosimetry studies in humans are now becoming possible, but they must be validated and their limitations defined. In addition, protein adducts, such as those found in sperm protamine and hemoglobin, are apparently stable for the lifetime of the cell, accurately indicate recent exposure, and should be considered in the estimation of genetic or carcinogenic risk whenever they can be correlated with DNA binding. Monitoring protein adducts has generally been considered to be a good surrogate procedure for measuring DNA-adduct formation in the target organ, but this should be validated in laboratory animals for each compound of interest.

REFERENCES

Aaron, C. S., and W. R. Lee. 1978. Molecular dosimetry of the mutagen ethyl methanesulfonate in *Drosophila melanogaster* spermatozoa: Linear relation of DNA alkylation per sperm cell (dose) to sex-linked recessive lethals. Mutat. Res. 49:27–44.

Abbott, P. J., and R. Saffhill. 1979. DNA synthesis with methylated poly (dC–dG) templates: Evidence for a competitive nature to miscoding by O^6-methylguanine. Biochim. Biophys. Acta. 562:51–61.

Adriaenssens, P. I., C. M. White, and M. W. Anderson. 1983. Dose-response relationships for the binding of benzo(a)pyrene metabolites to DNA and protein in lung, liver, and forestomach of control and butylated hydroxyanisole-treated mice. Cancer Res. 43:3712–3719.

Anderson, M. W. 1987. Carcinogen–DNA adducts as a measure of biological dose for risk analysis of carcinogenic data. Pp. 221–228 in National Research Council, Pharmacokinetics in Risk Assessment. Drinking Water and Health, Vol. 8. Washington, D.C.: National Academy Press.

Anderson, M. W., M. Boroujerdi, and A. G. E. Wilson. 1981. Inhibition *in vivo* of the formation of adducts between metabolites of benzo(a)pyrene and DNA by butylated hydroxyanisole. Cancer Res. 41:4309–4315.

Appleton, B. S., M. P. Goetchius, and T. C. Campbell. 1982. Linear dose-response curve for the hepatic macromolecular binding of aflatoxin B_1 in rats at very low exposures. Cancer Res. 42:3659–3662.

Arce, G. T., J. W. Allen, C. L. Doerr, E. Elmore, G. G. Hatch, M. M. Moore, Y. Sarief, D. Grunberger, and S. Nesnow. 1987. Relationships between benzo(a)pyrene–DNA adduct levels and genotoxic effects in mammalian cells. Cancer Res. 47:3388–3395.

Ashurst, S. W., G. M. Cohen, S. Nesnow, J. DiGiovanni, and T. J. Slaga. 1983. Formation of benzo(a)pyrene/DNA adducts and their relationship to tumor initiation in mouse epidermis. Cancer Res. 43:1024–1029.

Bannon, P., and W. Verly. 1972. Alkylation of phosphates and stability of phosphate triesters in DNA. Eur. J. Biochem. 31:103–111.

Bedell, M. A., J. G. Lewis, K. C. Billings, and J. A. Swenberg. 1982. Cell specificity in hepatocarcinogenesis: Preferential accumulation of O^6-methylguanine in target cell DNA during continuous exposure of rats to 1,2-dimethylhydrazine. Cancer Res. 42:3079–3083.

Beland, F. A., and F. F. Kadlubar. 1985. Formation and persistence of arylamine DNA adducts *in vivo*. Environ. Health Perspect. 62:19–30.

Beland, F. A., K. L. Dooley, and C. D. Jackson. 1982. Persistence of DNA adducts in rat liver and kidney after multiple doses of the carcinogen N-hydroxy-2-acetylaminofluorene. Cancer Res. 42:1348–1354.

Beland, F. A., N. F. Fullerton, T. Kinouchi, and M. C. Poirier. 1988. DNA adduct formation during continuous feeding of 2-acetylaminofluorene at multiple concentrations. Pp. 175–180 in Methods for Detecting DNA Damaging Agents in Humans: Applications in Cancer Epidemiology and Prevention, H. Bartsch, K. Hemminki, and I. K. O'Neill, eds. IARC Scientific Publications No. 89. Lyon: International Agency for Research on Cancer.

Belinsky, S. A., C. M. White, J. A. Boucheron, F. C. Richardson, J. A. Swenberg, and M. Anderson. 1986. Accumulation and persistence of DNA adducts in respiratory tissue of rats following multiple administrations of the tobacco specific carcinogen 4-(N-methyl-N-nitrosamino)-1-(3-pyridyl)-1-butanone. Cancer Res. 46:1280–1284.

Belinsky, S. A., C. M. White, T. R. Devereux, J. A. Swenberg, and M. W. Anderson. 1987. Cell selective alkylation of DNA in rat lung following low dose exposure to the tobacco specific carcinogen 4-(N-methyl-N-nitrosamino)-1-(3-pyridyl)-1-butanone. Cancer Res. 47:1143–1148.

Benzer, S. 1961. On the topography of the genetic fine structure. Proc. Natl. Acad. Sci. USA 47:403–415.

Beranek, D. T., R. H. Heflich, R. L. Kodell, S. M. Morris, and D. A. Casciano. 1983. Correlation between specific DNA-methylation products and mutation induction at the HGPRT locus in Chinese hamster ovary cells. Mutat. Res. 110:171–180.

Boucheron, J. A., F. C. Richardson, P. H. Morgan, and J. A. Swenberg. 1987. Molecular dosimetry of O^4-ethyldeoxythymidine in rats continuously exposed to diethylnitrosamine. Cancer Res. 47:1577–1581.

Branstetter, D. G., G. D. Stoner, H. A. J. Schut, D. Senitzer, P. B. Conran, and P. J. Goldblatt. 1987. Ethylnitrosourea-induced transplacental carcinogenesis in the mouse: Tumor response, DNA binding, and adduct formation. Cancer Res. 47:348–352.

Brookes, P. 1977. Mutagenicity of polycyclic aromatic hydrocarbons. Mutat. Res. 39:257–284.

Brookes, P., and P. D. Lawley. 1961. The reaction of mono- and di-functional alkylating agents with nucleic acids. Biochim. J. 80:496–503.

Brown, D. M. 1974. Chemical reactions of polynucleotides and nucleic acids. Pp. 1–90 in Basic Principles in Nucleic Acid Chemistry, Vol. II, P. O. P. Ts'O, ed. New York: Academic Press.

Cairns, J., P. Robins, B. Sedgwick, and P. Talmud. 1981. The inducible repair of alkylated DNA. Progr. Nucl. Acid Res. Mol. Biol. 26:237–244.

Casanova-Schmitz, M., T. B. Starr, and H. d'A. Heck. 1984. Differentiation between metabolic incorporation and covalent binding in the labeling of macromolecules in the rat nasal mucosa and bone marrow by inhaled [^{14}C]- and [^{3}H]formaldehyde. Toxicol. Appl. Pharmacol. 76:26–44.

Cohen, G. M., W. M. Bracken, R. P. Iyer, D. L. Berry, J. K. Selkirk and T. J. Slaga. 1979. Anticarcinogenic effects of 2,3,7,8-tetrachlorodibenzo-p-dioxin on benzo(a)pyrene and 7,12-dimethylbenz(a)-anthracene tumor initiation and its relationship to DNA binding. Cancer Res. 39:4027–4033.

Croy, R. G., and G. N. Wogan. 1981. Temporal patterns of covalent DNA adducts in rat liver after single and multiple doses of aflatoxin B_1. Cancer Res. 41:197–203.

Degen, G. H., and H. G. Neumann. 1981. Differences in aflatoxin B_1-susceptibility of rat and mouse are correlated with the capability in vitro to inactivate aflatox B_1-epoxide. Carcinogenesis 2:299–306.

Delclos, K. B., D. W. Miller, J. O. Lay, Jr., D. A. Casciano, R. P. Walker, P. P. Fu, and F. F. Kadlubar. 1987. Identification of C8-modified deoxyinosine and N^2-and C8-modified deoxyguanosine as major products of the in vitro reaction of N-hydroxy-6-aminochrysene with DNA and the formation of these adducts in isolated rat hepatocytes treated with 6-nitrochrysene and 6-aminochrysene. Carcinogenesis 8:1703–1709.

Dodson, L. A., R. S. Foote, S. Mitra, and W. E. Masker. 1982. Mutagenesis of bacteriophage T7 in vitro by incorporation of O^6-methylguanine during DNA synthesis. Proc. Natl. Acad. Sci. USA 79:7440–7444.

Drobetsky, E. A., A. J. Grosovsky, and B. W. Glickman. 1987. The specificity of UV-induced mutations at an endogenous locus in mammalian cells. Proc. Natl. Acad. Sci. USA 84:9103–9107.

Dunn, B. P. 1983. Wide-range linear dose-response curve for DNA binding of orally administered benzo(a)pyrene in mice. Cancer Res. 43:2654–2658.

Dyroff, M. C., F. C. Richardson, J. A. Popp, M. A. Bedell, and J. A. Swenberg. 1986. Correlation of O^4-ethyldeoxythymidine accumulation, hepatic initiation and hepatocellular carcinoma induction in rats continuously administered diethylnitrosamine. Carcinogenesis 7:241–246.

Ehrenberg, L., and S. Osterman-Golkar. 1980. Alkylation of macromolecules for detecting mutagenic agents. Teratogenesis Carcinog. Mutagen. 1:105–127.

Ehrenberg, L., S. Osterman-Golkar, D. Segerbäck, K. Svensson, and C. J. Calleman. 1977. Evaluation of genetic risks of alkylating agents. III. Alkylation of haemoglobin after metabolic conversion of ethene to ethene oxide in vivo. Mutat. Res. 45:175–184.

Everson, R. B., E. Randerath, R. M. Santella, R. C. Cefalo, T. A. Avitts, and K. Randerath. 1986. Detection of smoking-related covalent DNA-adducts in human placenta. Science 231:54–57.

Fahl, W. E., D. G. Scarpelli, and K. Gill. 1981. Relationship between benzo(a)pyreneinduced DNA base modification and frequency of reverse mutations in mutant strains of *Salmonella typhimurium*. Cancer Res. 41:3400–3406.

Frei, J. V., D. H. Swenson, W. Warren, and P. D. Lawley. 1978. Alkylation of deoxyribonucleic acid in vivo in various organs of C57Bl mice by the carcinogens *N*-methyl-*N*-nitrosourea, *N*-ethyl-*N*-nitrosourea and ethyl methanesulphonate in relation to induction of thymic lymphoma. Biochim. J. 174:1031–1044.

Frieberg, E. C. 1985. DNA Repair. New York: W. H. Freeman.

Gerchman, L. L., and D. B. Ludlum. 1973. The properties of O^6-methylguanine in templates for RNA polymerase. Biochim. Biophys. Acta. 308:310–316.

Goth, R., and M. F. Rajewsky. 1974. Persistence of O^6-ethylguanine in rat brain DNA: Correlation with nervous system-specific carcinogenesis by ethylnitrosourea. Proc. Natl. Acad. Sci. USA 71:639–643.

Green, C. L., E. L. Loechler, K. W. Fowler, and J. M. Essigmann. 1984. Construction and characterization of extrachromosomal probes for mutagenesis by carcinogens: Site-specific incorporation of O^6-methylguanine in viral and plasmid genomes. Proc. Natl. Acad. Sci. USA 81:13–17.

Harris, C. C., B. F. Trump, R. Grafstrom, and H. Autrup. 1982. Differences in metabolism of chemical carcinogens in cultured human epithelial tissues and cells. J. Cell. Biochem. 18:285–294.

Harris, C. C., K. Vahakangas, M. J. Newman, G. E. Trivers, A. Shamsuddin, N. Sinopoli, D. L. Mann, and W. E. Wright. 1985. Detection of benzo(a)pyrene diol epoxide-DNA adducts in peripheral blood lymphocytes and antibodies to the adducts in serum from coke oven workers. Proc. Natl. Acad. Sci. USA 82:6672–6676.

Haugen, A., G. Becher, C. Benestad, K. Vahakangas, G. E. Trivers, M. J. Newman, and C. C. Harris. 1986. Determination of polycyclic aromatic hydrocarbons in the urine, benzo(a)pyrene diol epoxide-DNA adducts in lymphocyte DNA, and antibodies to the adducts in sera from coke oven workers exposed to measured amounts of polycyclic aromatic hydrocarbons in the work atmosphere. Cancer Res. 46:4178–4183.

Heflich, R. H., S. M. Morris, D. T. Beranek, L. J. McGarrity, J. J. Chen, and F. A. Beland. 1986. Relationships between the DNA adducts and the mutations and sister-chromatid exchanges produced in Chinese hamster ovary cells by *N*-hydroxy-2-aminofluorene, *N*-hydroxy-N^1-acetyl benzidine 1-nitrosopyrine. Mutagenesis 1:201–206.

Hemminki, K. 1983. Nucleic acid adducts of chemical carcinogens and mutagens. Arch. Toxicol. 52:249–285.

Hoel, D. G., N. L. Kaplan, and M. W. Anderson. 1983. Implication of nonlinear kinetics on risk estimation in carcinogenesis. Science 219:1032–1037.

Kan, L. S., J. C. Barrett, P. S. Miller, and P. O. P. Ts'O. 1973. Proton magnetic resonance studies of the conformational changes of dideoxynucleoside ethyl phosphotriesters. Biopolymers 12:2225–2240.

Kensler, T. W., P. A. Egner, M. A. Trush, E. Bueding, and J. D. Groopman. 1985. Modification of aflatoxin B_1 binding to DNA *in vivo* in rats fed phenolic antioxidants, ethoxyquin and a dithiothione. Carcinogenesis 6:759–763.

Kensler, T. W., P. A. Egner, N. E. Davidson, B. D. Roebuck, A. Pikul, and J. D. Groopman. 1986. Modulation of aflatoxin metabolism, aflatoxin-N^7-guanine formation, and hepatic tumorigenesis in rats fed ethoxyquin: Role of induction of glutathione-*S*-transferases. Cancer Res. 46:3924–3931.

Kleihues, P., and G. P. Margison. 1974. Carcinogenicity of N-methyl-N-nitrosourea: Possible role of excision repair of O^6-methylguanine from DNA. J. Natl. Cancer Inst. 53:1839–1841.

Kleihues, P., and M. F. Rajewsky. 1984. Chemical neurooncogenesis: Role of structural DNA modifications, DNA repair and neural target cell population. Prog. Exp. Tumor Res. 27:1–16.

Kleihues, P., K. Patzschke, G. P. Margison, L. W. Wegner, and C. Mende. 1974. Reaction of methyl methanesulphonate with nucleic acids of fetal and newborn rats in vivo. Z. Krebsforsch. 81:273–283.

Kriek, E., and J. G. Westra. 1979. Metabolic activation of aromatic amines and amides and interaction with nucleic acids. Pp. 1–28 in Chemical Carcinogens and DNA, Vol II, P. L. Grover, ed. Boca Raton, Fla.: CRC Press.

Kroger, M., and B. Singer. 1979. Ambiguity and transcriptional errors as a result of methylation of the N-1 of purines and N-3 of pyrimidines. Biochemistry 18:3493–3500.

Lawley, P. D. 1974. Alkylation of nucleic acids and mutagenesis. Pp. 17–33 in Molecular and Environmental Aspects of Mutagenesis, L. Prakash, F. Sherman, M. W. Miller, C. W. Lawrence, and H. W. Taber, eds. 6th Rochester International Conference on Environmental Toxicity. Springfield, Ill.: Charles C Thomas.

Lawley, P. D., and P. Brookes. 1963. Further studies on the alkylation of nucleic acids and their constituent nucleotides. Biochim. J. 89:127–138.

Lawley, P. D., and C. N. Martin. 1975. Molecular mechanisms in alkylation mutagenesis: Induced reversion of bacteriophage T4rII AP72 by ethyl methanesulphonate in relation to extent and mode of ethylation of purines in bacteriophage deoxyribonucleic acid. Biochim. J. 145:85–91.

Leopold, W. R., E. C. Miller, and J. A. Miller. 1979. Carcinogenicity of antitumor cis-platinum(II) coordination complexes in the mouse and rat. Cancer Res. 39:913–918.

Lewis, J. G., and J. A. Swenberg. 1980. Differential repair of O^6-methylguanine in DNA of rat hepatocytes and non-parenchymal cells. Nature 288:185–187.

Lewis, J. G., and J. A. Swenberg. 1983. The kinetics of DNA alkylation, repair and replication in hepatocytes, Kupffer cells, and sinusoidal endothelial cells in rat liver during continuous exposure to 1,2-dimethylhydrazine. Carcinogenesis 4:529–536.

Lindamood, C., III, M. A. Bedell, K. C. Billings, and J. A. Swenberg. 1982. Alkylation and de novo synthesis of liver cell DNA from C_3H mice during continuous dimethylnitrosamine exposure. Cancer Res. 42:4153–4157.

Loechler, E. L., C. L. Green, and J. M. Essigmann. 1984. In vivo mutagenesis by O^6-methylguanine built into a unique site in a viral genome. Proc. Natl. Acad. Sci. USA 81:6271–6275.

Lutz, W. K. 1979. In vivo covalent binding of organic chemicals to DNA as a quantitative indicator in the process of chemical carcinogenesis. Mutat. Res. 65:289–356.

McCormick, J. J., and V. M. Maher. 1985. Cytotxic and mutagenic effects of specific carcinogen-DNA adducts in diploid human fibroblasts. Environ. Health Perspect. 62:145–155.

Miller, E. C. 1978. Some current perspectives on chemical carcinogenesis in humans and experimental animals: Presidential address. Cancer Res. 38:1479–1496.

Miller, P. S., K. N. Fang, N. S. Kondo, and P. O. P. Ts'O. 1971. Synthesis and properties of adenine and thymine nucleoside alkyl phosphotriesters, the neutral analogs of dinucleoside monophosphates. J. Am. Chem. Soc. 93:6657–6665.

Miller, P. S., J. C. Barrett, and P. O. P. Ts'O. 1974. Synthesis of oligodeoxyribonucleotide ethyl phosphotriesters and their specific complex formation with transfer ribonucleic acid. Biochemistry 13:4887–4896.

Monroe, D. H., and D. L. Eaton. 1987. Comparative effects of butylated hydroxyanisole on hepatic in vivo DNA binding in vitro biotransformation of aflatoxin B_1 in the rat and mouse. Toxicol. Appl. Pharmacol. 90:401–409.

Neumann, H.-G. 1983. Role of extent and persistence of DNA modifications in chemical carcinogenesis by aromatic amines. Recent Results Cancer Res. 84:77–89.

Neumann, H.-G. 1984. Analysis of hemoglobin as a dose monitor for alkylating and arylating agents. Arch. Toxicol. 56:1–6.

Neumann, H.-G., H. Baur, and R. Wirsing. 1980. Dose relationship in the primary lesion of strong electrophilic carcinogens. Arch. Toxicol. 3(Suppl.):69–77.

Newbold, R. F., P. Brookes, and R. G. Harvey. 1979. A quantitative comparison of the mutagenicity of carcinogenic polycyclic hydrocarbons derivates in cultured mammalian cells. Int. J. Cancer 24:203–209.

Newbold, R. F., W. Warren, A. S. C. Metcalf, and J. Amos. 1980. Mutagenicity of carcinogenic methylating agents is associated with a specific DNA modification. Nature 283:596–599.

NRC (National Research Council). 1973. Toxicants Occurring Naturally in Foods, 2nd ed. Washington, D.C.: National Academy of Sciences. 624 pp.

NRC (National Research Council). 1983. Risk Assessment in the Federal Government: Managing the Process. Washington, D.C.: National Academy Press. 191 pp.

NRC (National Research Council). 1986. Drinking Water and Health, Vol. 6. Washington, D.C.: National Academy Press. 457 pp.

NRC (National Research Council). Committee on Biological Markers. 1987. Biological markers in environmental health research. Environ. Health Perspect. 74:3–9.

O'Connor, P. J., M. J. Capps, and A. W. Craig. 1973. Comparative studies of the hepatocarcinogen N,N-dimethylnitrosamine in vivo: Reaction sites in rat liver DNA and the significance of their relative stabilities. Br. J. Cancer 27:153–166.

O'Connor, P. J., G. P. Margison, and A. W. Craig. 1975. Phosphotriesters in rat liver deoxyribonucleic acid after the administration of the carcinogen N,N-dimethylnitrosamine in vivo. Biochem. J. 145:475–482.

Osterman-Golkar, S., P. B. Farmer, D. Segerbck, E. Bailey, C. J. Calleman, K. Svensson, and L. Ehrenberg. 1983. Dosimetry of ethylene oxide in the rat by quantitation of alkylated histidine in hemoglobin. Teratogenesis Carcinog. Mutagen. 3:395–405.

Osterman-Golkar, S. L., L. Ehrenberg, D. Segerbäck, and I. Hällstrom. 1976. Evaluation of genetic risks of alkylating agents. II. Haemoglobin as a dose monitor. Mutat. Res. 34:1–10.

Pegg, A. E. 1983. Alkylation and subsequent repair of DNA after exposure to dimethylnitrosamine and related carcinogens. Rev. Biochem. Toxicol. 5:83–133.

Pereira, M. A., and L. W. Chang. 1981. Binding of chemical carcinogens and mutagens to rat hemoglobin. Chem.-Biol. Interact. 33:301–305.

Pereira, M. A., and L. W. Chang. 1982. Binding of chloroform to mouse and rat hemoglobin. Chem.-Biol. Interact. 39:89–99.

Pereira, M. A., F. J. Burns, and R. E. Albert. 1979. Dose response for benzo(a)pyrene in mouse epidermal DNA. Cancer Res. 39:2556–2559.

Pereira, M. A., L.-H. C. Lin, and L. W. Chang. 1981. Dose dependency of 2-acetylaminofluorene binding to liver DNA and hemoglobin in mice and rats. Toxicol. Appl. Pharmacol. 60:472–478.

Perera, F. P., M. C. Poirier, S. H. Yuspa, J. Nakayama, A. Jaretzki, M. M. Curnen, D. M. Knowles, and I. B. Weinstein. 1982. A pilot project in molecular cancer epidemiology: Determination of benzo[a]pyrene-DNA adducts in animal and human tissues by immunoassays. Carcinogenesis 3:1405–1410.

Perera, F. P., R. M. Santella, D. Brenner, M. C. Poirier, A. A. Munshi, H. K. Fischman, and J. Van Ryzin. 1987a. DNA adducts, protein adducts, and sister chromatid exchange in cigarette smokers and nonsmokers. J. Natl. Cancer Inst. 79:449–456.

Perera, F. P., K. Hemminki, R. M. Santella, D. Brenner, and G. Kelly. 1987b. DNA adducts in white blood cells of foundry workers (Meeting abstract). Proc. Annu. Meet. Am. Assoc. Cancer Res. 28:94.

Phillips, D. H., K. Hemminki, A. Alhonen, A. Hewer, and P. L. Grover. 1988. Monitoring occupational exposure to carcinogens: Detection by ^{32}P-postlabeling of aromatic DNA adducts in white blood cells from iron foundry workers. Mutat. Res. 204:531–541.

Poirier, M. C., and F. A. Beland. 1987. Determination of carcinogen-induced macromolecular adducts in animals and humans. Prog. Exp. Tumor Res. 31:1–10.

Poirier, M. C., J. M. Hunt, B. A. True, B. A. Laishes, J. F. Young, and F. A. Beland. 1984. DNA adduct formation, removal and persistence in rat liver during one month of feeding 2-acetylaminofluorene. Carcinogenesis 5:1591–1596.

Poirier, M., E. Reed, L. Zwelling, R. Ozols, C. Litterest, and S. Yuspa. 1985. Polyclonal antibodies to quantitate cis-diamminedichloroplatinum (II)–DNA-adducts in cancer patients and animal models. Environ. Health Perspect. 62:89–94.

Poirier, M. C., E. Reed, R. F. Ozols, T. Fasy, and S. H. Yuspa. 1987. DNA adducts of cisplatin in nucleated peripheral blood cells and tissues of cancer patients. Prog. Exp. Tumor Res. 31:104–113.

Rajewsky, M. F., L. H. Augenlicht, H. Biessmann, R. Goth, D. F. Huelser, O. D. Laerum, and L. Y. Lomakina. 1977. Nervous system-specific carcinogenesis by ethyl nitrosourea in the rat: Molecular and cellular aspects. Pp. 709–726 in Origins of Human Cancer, H. H. Hiatt, J. D. Watson, and J. A. Winsten, eds. New York: Cold Spring Harbor Laboratory.

Reddy, M. V., P. C. Kenny, and K. Randerath. 1987. ^{32}P-assay of DNA adducts in white blood cells (WBC) and placentas of pregnant women exposed to residential wood combustion (RWC) smoke (Meeting abstract). Proc. Annu. Meet. Am. Assoc. Cancer Res. 28:97.

Reed, E., R. F. Ozols, R. Tarone, S. H. Yuspa, and M. C. Poirier. 1987. Platinum-DNA adducts in leukocyte DNA correlate with disease response in ovarian cancer patients receiving platinum-based chemotherapy. Proc. Natl. Acad. Sci. USA 84:5024–5028.

Richardson, F. C., M. C. Dyroff, J. A. Boucheron, and J. A. Swenberg. 1985. Differential repair of O^4-alkylthymidine following exposure to methylating and ethylating hepatocarcinogens. Carcinogenesis 6:625–629.

Richardson, K. K., F. C. Richardson, R. M. Crosby, J. A. Swenberg, and T. R. Skopek. 1987. DNA base changes and alkylation following in vivo exposure of Escherichia coli to N-methyl-N-nitrosourea or N-ethyl-N-nitrosourea. Proc. Natl. Acad. Sci. USA 84:344–348.

Russell, W. L. 1984. Dose-response, repair and no-effect dose levels in mouse germ-cell mutagenesis. Pp. 153–160 in Problems of Threshold in Chemical Mutagenesis, Y. Tazima, S. Kondo, and Y. Kuroda, eds. Tokyo: Environmental Mutagen Society of Japan.

Russell, W. L., P. R. Hunsicker, D. A. Carpenter, C. V. Cornett, and G. M. Guinn. 1982. Effect of dose fractionation on the ethylnitrosourea induction of specific locus mutations in mouse spermatogonia. Proc. Natl. Acad. Sci. USA 79:3592–3593.

Saul, R. L., and B. N. Ames. 1986. Background levels of DNA damage in the population. Pp. 529–535 in Mechanisms of DNA Damage and Repair, M. G. Simic, L. Grossman, and A. C. Upton, eds. Basic Life Sciences, Vol. 38. New York: Plenum.

Sega, G. A., and J. G. Owens. 1978. Ethylation of DNA and protamine by ethyl methanesulfonate in the germ cells of male mice and the relevancy of these molecular targets to the induction of dominant lethals. Mutat. Res. 52:87–106.

Sega, G. A., and J. G. Owens. 1983. Methylation of DNA and protamine by methyl methanesulfonate in the germ cells of male mice. Mutat. Res. 111:227–244.

Sega, G. A., and J. G. Owens. 1987. Binding of ethylene oxide in spermiogenic germ cell stages of the mouse after low-level inhalation exposure. Environ. Molecul. Mutag. 10:119–127.

Sega, G. A., C. R. Rohrer, H. R. Harvey, and A. E. Jetton. 1986. Chemical dosimetry of ethyl nitrosourea in the mouse testis. Mutat. Res. 159:65–74.

Segerbäck, D., C. J. Calleman, L. Ehrenberg, G. Lofroth, and S. Osterman-Golkar. 1978. Evaluation of genetic risks of alkylating agents. IV. Quantitative determination of alkylated amino acids in haemoglobin as a measure of the dose after treatment of mice with methyl methanesulfonate. Mutat. Res. 49:71–82.

Setlow, R. B. 1983. Variations in DNA repair among humans. Pp. 231–254 in Human Carcinogenesis, C. C. Harris and H. N. Autrup, eds. New York: Academic Press.

Setlow, R. B. 1987. Theory presentation and background summary. Pp. 177–182 in Modern Biological Theories of Aging, H. R. Warner, R. N. Butler, R. L. Sprott, and E. L. Schneider, eds. New York: Raven.

Shamsuddin, A. K., N. T. Sinopoli, K. Hemminki, R. R. Boesch, and C. C. Harris. 1985. Detection of benzo(a)pyrene: DNA-adducts in human white blood cells. Cancer Res. 45:66–68.

Shelby, M. D., K. T. Cain, L. A. Hughes, P. W. Braden, and W. M. Generoso. 1986. Dominant lethal effects of acrylamide in male mice. Mutat. Res. 173:35–40.

Shelby, M. D., K. T. Cain, C. V. Cornett, and W. M. Generoso. 1987. Acrylamide: Induction of heritable translocations in male mice. Environ. Mutag. 9:363–368.

Shugart, L. 1985. Quantitating exposure to chemical carcinogens: In vivo alkylation of hemoglobin by benzo[a]pyrene. Toxicology 34:211–220.

Sims, P., and P. L. Grover. 1974. Epoxides in polycyclic aromatic hydrocarbon metabolism and carcinogenesis. Adv. Cancer Res. 20:165–274.

Singer, B. 1975. The chemical effects of nucleic acid alkylation and their relationship to mutagenesis and carcinogenesis. Prog. Nucleic Acid Res. Mol. Biol. 15:219–284.

Singer, B. 1982. Mutagenesis from a chemical perspective: Nucleic acid reactions, repair, translation, and transcription. Basic Life Sci. 20:1–42.

Singer, B. 1985. In vivo formation and persistence of modified nucleosides resulting from alkylating agents. Environ. Health Perspect. 62:41–48.

Singer, B., and H. Fraenkel-Conrat. 1969. The role of conformation in chemical mutagenesis. Prog. Nucleic Acid Res. Mol. Biol. 9:1–29.

Singer, B., and D. Grunberger. 1983. Molecular Biology of Mutagens and Carcinogens. New York: Plenum. 347 pp.

Singer, B., H. Fraenkel-Conrat, and J. T. Kusmierek, 1978a. Preparation and template activities of polynucleotides containing O^2- and O^4-alkyluridine. Proc. Natl. Acad. Sci. USA 75:1722–1726.

Singer, B., M. Kroger, and M. Carrano. 1978b. O^2- and O^4-alkyl-pyrimidine nucleosides: Stability of the gylcosyl bond and of the alkyl group as a function of pH. Biochemistry 17:1246–1250.

Singer, B., R. G. Pergolizzi, and D. Grunberger. 1979. Synthesis and coding properties of dinucleoside diphosphates containing alkyl pyrimidines which are formed by the action of carcinogens on nucleic acids. Nucleic Acids Res. 6:1709–1719.

Singer, B., J. Sági, and J. T. Kúsmierek. 1983a. Escherichia coli polymerase I can use O^2-methyldeoxythymidine or O^4-methyl-deoxythymidine in place of deoxythymidine in primed poly (dA-dT)•poly(dA-dT) synthesis. Proc. Natl. Acad. Sci. USA 80:4884–4888.

Singer, B., J. T. Kusmierek, and H. Fraenkel-Conrat. 1983b. In vitro discrimination of replicases acting on carcinogen-modified polynucleotide templates. Proc. Natl. Acad. Sci. USA 80:969–972.

Skopek, T. R., R. D. Wood, and F. Hutchinson. 1985. Sequence specificity of mutagenesis in the C^1 gene of bacteriophage lambda. Environ. Health Perspect. 62:157–161.

Snow, E. T., R. S. Foote, and S. Mitra. 1983. Replication and demethylation of O^6-methylguanine in DNA. Prog. Nucleic Acid Res. Mol. Biol. 29:99–103.

Staffa, J. A., and M. A. Mehlman, eds. 1979. Innovations in Cancer Risk Assessment (ED_{01} Study). Park Forest South, Ill.: Pathotox. 246 pp.

Stowers, S. J., and M. W. Anderson. 1985. Formation and persistence of benzo(a)pyrene metabolite–DNA adducts. Environ. Health Perspect. 62:31–39.

Sun, L., and B. Singer. 1975. The specificity of different classes of ethylating agents toward various sites of HeLa cell DNA in vitro and in vivo. Biochemistry 14:1795–1802.

Swenberg, J. A., and T. R. Fennell. 1987. DNA damage and repair in mouse liver. Arch. Toxicol. (Suppl.) 10:162–171.

Swenberg, J. A., M. A. Bedell, K. C. Billings, D. R. Umbenhauer, and A. E. Pegg. 1982. Cell specific differences in O6-alkylguanine DNA repair activity during continuous exposure to carcinogen. Proc. Natl. Acad. Sci. USA 79:5499–5502.

Swenberg, J. A., M. C. Dyroff, M. A. Bedell, J. A. Popp, N. Huh, A. Kirstein, and M. F. Rajewsky. 1984. O^4-ethyldeoxythymidine, but not O6-ethyldeoxyguanosine, accumulates in hepatocytes of DNA of rats exposed continuously to diethylnitrosamine. Proc. Natl. Acad. Sci. USA 81:1692–1695.

Swenberg, J. A., F. C. Richardson, J. A. Boucheron, and M. C. Dyroff. 1985. Relationships between DNA adduct formation and carcinogenesis. Environ. Health Perspect. 62:177–183.

Swenberg, J. A., F. C. Richardson, J. A. Boucheron, F. H. Deal, S. A. Belinsky, M. Charbonneau, and B. G. Short. 1987. High- to low-dose extrapolation: Critical determinants involved in the dose-response of carcinogenic substances. Environ. Health Perspect. 76:57–63.

Swenson, D. H., and P. D. Lawley. 1978. Alkylation of deoxyribonucleic acid by carcinogens dimethyl sulphate, ethyl methanesulphonate, N-ethyl-N-nitrosourea and N-methyl-N-nitrosourea. Biochim. J. 171:575–587.

Talaska, G., W. W. Au, J. B. Ward, Jr., K. Randerath, and M. S. Legator. 1987. The correlation between DNA adducts and chromosomal aberrations in the target organ of benzidine exposed, partially-hepatectomized mice. Carcinogenesis 8:1899–1905.

Tannenbaum, S. R., P. L. Skipper, L. C. Green, M. W. Obiedzinski, and F. Kadlubar. 1983. Blood protein adducts as monitors of exposure to 4-aminobiphenyl. Abstract 271. Proc. Am. Assoc. Cancer Res. 24:69.

Thilly, W. G. 1985. The potential use of gradient denaturing gel electrophoresis to obtain mutation spectra in human cells. Pp. 511–528 in The Role of Chemicals and Radiation in the Etiology of Cancer, E. Huberman, ed. Carcinogenesis, Vol. 10. New York: Raven Press.

Tice, R. B., and R. B. Setlow. 1985. DNA repair and replication in aging organisms and cells. Pp. 173–224 in Handbook of the Biology of Aging, 2nd ed, C. E. Finch and E. L. Schneider, eds. New York: Van Nostrand Reinhold.

Travis, C. C., R. K. White, A. D. Arms. 1989. A physiologically-based pharmacokinetic approach to assessing the cancer risk of tetrachloroethylene. Pp. 769–795 in The Risk Assessment of Environmental and Human Health Hazards: A Textbook of Case Studies, D. Paustenbach, ed. New York: John Wiley & Sons.

Tullis, D. L., K. L. Dooley, D. W. Miller, K. P. Baetcke, and F. F. Kadlubar. 1987. Characterization and properties of the DNA adducts formed from N-methyl-4-aminoazobenzene in rats during a carcinogenic treatment regimen. Carcinogenesis 8:577–583.

Umbenhauer, D., C. P. Wild, R. Montesano, R. Saffhill, J. M. Boyle, N. Huh, U. Kirstein, J. Thomale, M. F. Rajewsky, and S. H. Lu. 1985. O^6-methyldeoxyguanosine in oesophageal DNA among individuals at high risk of oesophageal cancer. Int. J. Cancer 36:661–665.

van Zeeland, A. A., G. R. Mohn, A. Neuhauser-Klaus, and U. H. Ehling. 1985. Quantitative comparison of genetic effects of ethylating agents on the basis of DNA adduct formation: Use of O^6-ethylguanine as molecular dosimeter for extrapolation from cells in culture to the mouse. Environ. Health Perspect. 62:163–169.

Vrieling, H., J. W. I. M. Simons, and A. A. van Zeeland. 1988. Nucleotide sequence determination of point mutations at the mouse HPRT locus using *in vitro* amplification of HPRT mRNA sequences. Mut. Res. 198:107–113.

Wild, C. P., R. C. Garner, R. Montesano, and F. Tursi. 1986. Aflatoxin B_1 binding to plasma albumin and liver DNA upon chronic administration to rats. Carcinogenesis 7:853–858.

Wogan, G. N. 1988. Detection of DNA damage in studies on cancer etiology and prevention. Pp. 32–54 in Methods for Detecting DNA Damaging Agents in Humans: Applications in Cancer Epidemiology and Prevention, H. Bartsch, K. Hemminki, and I. K. O'Neill, eds. IARC Scientific Publications No. 89. Lyon: International Agency for Research on Cancer.

Wogan, G. N., and N. J. Gorelick. 1985. Chemical and biochemical dosimetry of exposure to genotoxic chemicals. Environ. Health Perspect. 62:5–18.

Yang, L. L., V. M. Maher, and J. J. McCormick. 1980. Error-free excision of the cytotoxic, mutagenic N^2-deoxyguanosine DNA adduct formed in human fibroblasts by ($\pm$)-7$B^3$8a^2-dihydroxy-9a^2,10a^2-epoxy-7,8,9,10-tetrahydrobenzo(a)pyrene. Proc. Natl. Acad. Sci. USA 10:5933–5937.

2

DNA-Adduct Technology

The consensus of several recent meetings on DNA-adduct research has been that new assay systems for detecting and measuring DNA adducts and protein adducts have the potential to improve markedly the biologic bases for estimating the risks of human exposure to several important classes of environmental pollutants (Bartsch et al., 1988; Berlin et al., 1984; de Serres, 1988; de Serres et al., 1985; Farmer et al., 1987). This chapter describes and evaluates some of the currently used assays and discusses how the information they provide can lead to the improvement of risk assessment and epidemiological studies.

New DNA-adduct technology embodies striking technologic improvements in sensitivity and specificity and permits measurement of mammalian response to small, intermittent environmental exposures. For example, molecular dosimetric assays and mutation analysis in mammalian cells in tissue culture suggest that low exposures to genetic toxicants produce DNA lesions at approximately 2,000 per cell in the lung after exposure to aromatic amines and 100,000 per cell in the upper layer of skin after exposure to the ultraviolet component of sunlight (Lohman et al., 1985). Those figures correspond to about 1 adduct per 10^{-7} bases in the lung and at least 1 adduct per 10^{-4} bases per day in the upper layer of skin.

The new methods of measuring DNA adducts are in many instances relatively inexpensive, fast, and reproducible. They can be applied to readily available samples of body fluids, such as blood, semen, and urine, and to small samples of cells, such as buccal mucosa or skin biopsy specimens.

TECHNIQUES FOR DETECTING DNA ADDUCTS

Chromatographic and Spectrometric Methods

Chromatography has many variations, but all involve the flow of test material through tubing containing stationary material designed to adsorb the components of the test mixture selectively, and thus create different flow rates and a series of bands (chromatograms) by which their identity can be determined.

In spectrometry, the sample interacts with light or particles to yield distinctive spectral signals. One version is mass spectrometry, the most accurate technique for trace organic-chemical analysis, according to the National Bureau of Standards.

The polarity and size of DNA adducts, or fragments of DNA adducts, can markedly influence the separation power and thus the sensitivity of these techniques. Each analytic method possesses potentially good to high resolving power, but the suitability and sensitivity of any method or combination of methods commonly depends on the physicochemical properties of the adduct or the class of adducts to be tested.

LIQUID CHROMATOGRAPHY/MASS SPECTROMETRY (LC/MS)

In this method, DNA adducts are separated from the test sample by adsorption on an activated surface. They are then put into a mass spectrometer for final analysis.

HIGH-PERFORMANCE LIQUID CHROMATOGRAPHY (HPLC)

This method combines speed and high-resolution power to fractionate DNA adducts by column chromatography of modified DNA bases. It can detect fluorescence.

ATOMIC-ABSORPTION SPECTROMETRY (AAS)

This quantitative analytic technique detects adducts containing metals by absorbing light of specific wavelengths from excited atoms.

TANDEM MASS SPECTROMETRY (MS/MS)

Two mass spectrometers are used in sequence to detect fragments (ions) of specific molecules, such as DNA adducts. An ion from the mixture is selected and focused through the first mass spectrometer; thereafter, it is fragmented into smaller portions for further, high-resolution analysis in the second mass spectrometer (Farmer et al., 1988).

FLUORESCENCE LINE-NARROWING SPECTROMETRY (FLNS)

Particularly useful for fluorescent adducts of polycyclic aromatic hydrocarbons, FLNS uses low temperature and laser excitation to improve sensitivity in the quantitative fluorometric analysis of DNA adducts (Jankowiak et al., 1988).

ULTRAVIOLET RADIATION/HIGH-PERFORMANCE LIQUID CHROMATOGRAPHY (UV/HPLC)

Because all adducts absorb ultraviolet radiation, testing for UV absorption is useful for large, bulky adducts, such as those formed by aflatoxin, but is rarely sensitive enough for human monitoring. Sensitive HPLC can be added to detect fluorescence after hydrolysis of particular carcinogen–DNA adducts. For example, synchronous fluorescence spectrometry (SFS) scans a sample with a fixed wavelength difference between excitation and emission. Three-dimensional plots of fluorescence intensity, emission, and excitation-emission wavelength difference should be able to help identify some unknown adducts (Farmer et al., 1987).

GAS CHROMATOGRAPHY/ELECTRON CAPTURE NEGATIVE ION MASS SPECTROMETRY (GC/ECNIMS)

In this technique, the test material is derivatized and volatilized into a carrier gas stream whose components pass through a chromatography column at different rates, where the adducts are separated. The sample then enters the mass spectrometer, where distinctive ions are formed and detected with high sensitivity and specificity. A related, less specific method is GC with electron capture detection (GC/ECD).

Quantitative Immunoassays

Monoclonal or polyclonal antisera specific for carcinogen–DNA adducts or carcinogen-modified DNA are used in immunoassays to quantify the binding of known carcinogens with DNA in biologic samples of nucleic acid (Poirier, 1984). These immunoassays are simple to perform, inexpensive, and thus appropriate for human samples; but the antisera are chemical-specific, and thus different antisera must be developed for each adduct of interest (Santella, 1988).

The most sensitive immunoassays are run in a competitive mode, in which two chemically identical haptens (in this case, DNA adducts) compete for an antibody binding site. The concentration of 1 hapten (usually an assay standard) is always kept constant, and that hapten is radioactively labeled (in radioimmunoassays) or bound to the bottom of microtiter wells (in enzyme-linked immunosorbent assays). The other hapten is used in increasing concentrations to compete with the constant hapten for binding to the antibody. The variable hapten can be either a standard immunogen or an unknown sample. Quantitation is based on comparison of unknowns with the inhibition curve generated by the standard immunogen.

RADIOIMMUNOASSAY (RIA)

In conventional RIA, a competitive technique, the antigen–antibody complex is separated from the whole mixture in tubes by a variety of physical or chemical methods (Poirier, 1981), and the standard hapten is radioactively labeled.

ENZYME-LINKED IMMUNOSORBENT ASSAY (ELISA)

In ELISA a solid-phase competitive assay that uses microtiter plates, the antibody bound after competition can be measured by a second enzyme-linked antibody used to cleave a specific substrate. One of the most commonly used enzyme conjugates is alkaline phosphatase, which cleaves a variety of phosphorylated substrates into products that can be detected by spectrophotometry or fluorescence (ELISA) (Poirier, 1981). In general, microtiter-plate assays can be more sensitive than RIAs, but they are also more often inconsistent and variable.

ULTRASENSITIVE ENZYMATIC RADIOIMMUNOASSAY (USERIA)

The procedure for USERIA is similar to that of ELISA, except that the substrate obtained after application of the first antibody is labeled with radioactive isotopes for radiochemical measurement of enzyme interaction. Counting requires that samples be manually removed from each well (Farmer et al., 1987; Santella, 1988).

Immunohistochemical Techniques

Enzyme-staining with a fluorescence or peroxidase end point can identify the specific cell types in which DNA adducts occur (and thus the cells that are targets for carcinogens) within a complex tissue sample. Monoclonal or polyclonal antibodies are used in conjunction with other antibodies that contain peroxidase enzyme or a fluorescent probe. With the fluorescent probe, images of the adducts can be enhanced by computer for analysis. Polyclonal antibodies are less easily identified by the most sensitive image analysis systems (Adamkiewicz et al., 1985), but they have been used with microfluorometry for semiquantitative comparison between samples (Huitfeldt et al., 1987) and to detect, but not quantify, binding in human tissues (den Engelse et al., 1988).

^{32}P-Postlabeling Technique

DNA is enzymatically hydrolyzed and the digest is labeled with phosphorylating enzyme to incorporate radioactivity. Thin-layer chromatography (TLC) is used to separate the adducts, which can be detected by autoradiography, and a portion of the chromatogram is excised for estimation of the total count (Gupta et al., 1982). Less useful for low-molecular-weight compounds, ^{32}P-postlabeling is more sensitive for aromatic or bulky hydrophobic adducts and has been able to show the extent and persistence of adduct formation in animals by more than 70 compounds, including aromatic hydrocarbons, aromatic amines, estrogens, and methylating agents (Gupta and Randerath, 1988). With ^{32}P-postlabeling, it is possible to monitor human carcinogen exposure (Randerath et al., 1988). This technique is also capable of analyzing DNA adducts formed by unknown hydrophobic compounds (Farmer et al., 1987); tandem technology might someday be developed for identifying the structure of such adducts.

TECHNIQUES FOR DETECTING PROTEIN ADDUCTS

Some of the same methods for detecting DNA adducts are applied to the determination of protein adducts (GC, GC–MS, immunoassay, and fluorescence detection with HPLC). New techniques for detecting protein adducts, especially those using hemoglobin as a target molecule (Bailey et al., 1987; Osterman-Golkar, 1988; Neumann, 1984, 1988), offer high sensitivity and specificity in detecting exposure of animals to alkylating agents. Several studies have found a direct correlation between hemoglobin-adduct and DNA-adduct concentrations in exposed experimental animals (Adriaenssens et al., 1983; Pereira, et al., 1981; Shugart, 1985; Wild et al., 1986). An additional practical advantage of measuring protein adducts is that large amounts of some proteins (especially hemoglobin) can be obtained from human subjects. Thus, protein adducts might provide more reliable measurements than DNA adducts for evaluating both normal background concentrations of adducts and deviations from the normal.

SENSITIVITY AND SPECIFICITY

The intrinsic sensitivity of an assay to detect the molecular effect of a given hypothetical chemical is usually expressed as femtomoles (fmol; 10^{-15} moles) of adduct per milligram of DNA or protein. Tables 2-1 and 2-2 give estimates of the intrinsic sensitivities of various DNA and protein binding assays. The estimates are approximations and sensitivity might vary by several orders of magnitude, depending on the physicochemical nature of the chemical or adduct.

Tables 2-1 and 2-2 show that some of the immunochemical assays and the postlabeling assay have good sensitivity and do not require invasive techniques or large tissue samples. The MS/MS method, a physicochemical method, has the same advantages as the immunochemical and postlabeling assays, but its dependence on expensive and sophisticated equipment could severely limit widespread application. Other techniques, such as immunochemical methods for detecting DNA adducts at the single-cell level (Baan et al., 1988; Perera et al., 1988; Van Benthem et al., 1988) and the recently introduced laser-scan immunofluorescence microscopy (Baan et al., 1986), are also limited by unique instrumentation requirements. However, the MS/MS (Farmer et al., 1988), RIA (Umbenhauer et al., 1985), and ^{32}P-postlabeling (Randerath et al., 1988) methods can detect interactions with small amounts of unidentified alkylating agents associated with occupational and low-level environmental exposures.

Both physicochemical and immunochemical methods for detecting adducts or metabolites of genetic toxicants in urine have high sensitivity and specificity (Oshima and Bartsch, 1988; Shuker and Farmer, 1988; Vanderlaan et al., 1988), but still require validation as indicators of internal exposure. A high adduct concentration in urine is often assumed to indicate high internal exposure, but this assumption is not necessarily correct. Proper mass-balance evaluations are needed to measure the intake and excretion of genetically toxic agents and their metabolites. Without such determinations, it is equally justifiable to relate the presence of a high concentration of adducts in urine to detoxification or to low internal exposure (Lohman et al., 1984; van Sittert, 1984).

Although technologic improvements make feasible the sensitive measurement of exposure to genetic toxicants in animal models and humans, no generally applicable methods have been developed for estimating genetic risk (Wogan, 1988). Attempts are under way to relate target dose in humans to biologically adverse effects of small exposures to genetically toxic agents (Ehrenberg, 1988).

The use of these new, ultrasensitive analytical techniques in risk assessment will depend on an understanding of the mechanistic relationships between DNA alterations and the ultimate expression of toxic effects. Recent devel-

TABLE 2-1 Assays for Detection of Interaction Between Genetically Toxic Agents and DNA

Method for Detecting DNA Adducts[a]	Estimated Lower Limit of Detection, fmol/mg of DNA[b]	DNA Used per Assay, μg[c]	Advantages	Disadvantages	References
UV/HPLC	30,000	1,000	Easy detection	Very low sensitivity	Lohman, 1988, personal communication[f]
AAS	2	50	Easy detection	Useful only for metal-containing compounds	Knox et al., 1986
FL/HPLC	300	100	Simple, routine application possible	Analyte must be fluorescent	Rahn et al., 1982
FLNS	30	100	Direct, nondestructive	Analyte must be fluorescent, sophisticated equipment needed	Heisig et al., 1984; Sanders et al., 1986
LC/MS	30	500	Multiple adducts detectable	Analyte must be (made) volatile	Giese, 1988, personal communication[e]
GC/ECNIMS or GC/ECD	3[d]	500	Multiple adducts detectable	Analyte must be (made) volatile	Giese, 1986; Minnetian et al., 1987; Adams et al., 1986; Mohamed et al., 1984
MS/MS	3	10	Multiple adducts detectable	Sophisticated, expensive equipment	Farmer et al., 1988
Immunochemical: Direct	30.0	0.002	Low costs, routine application possible	Limited availability of antibodies, quantitation difficult	Harris et al., 1982; Muller and Rajewsky, 1981; Poirier, 1984; Santella et al., 1987

Competitive	100	50	Low costs, routine application possible	Limited availability of antibodies, quantitation difficult	Harris et al., 1982; Muller and Rajewsky, 1981; Poirier, 1984; Santella et al., 1987
Single cell:					
ES	3,000	100 cells or tissue section	Applicable to identified cell types	Limited availability of antibodies; technically complex; sophisticated equipment	Lohman, 1988, personal communication[f]
FL	30	100 cells or tissue section	Applicable to identified cell types	Limited availability of antibodies; technically complex; sophisticated equipment	Huitfeldt et al., 1986
^{32}P-postlabeling	0.3	10	Low costs, multiple adducts detectable, routine application possible	Radioisotopes needed; adducts lack chemical characterization; time-consuming; technically complex; up to 10^4 times less sensitive for small adducts	Gupta and Randerath, 1988

[a] Abbreviations: UV, 254-nm ultraviolet radiation; HPLC, high-performance liquid chromatography; AAS, atomic-absorption spectrometry; FL, immunofluorescence microscopy; FLNS, fluorescence line narrowing spectrometry; LC/MS, liquid chromatography/mass spectrometry; GC/ECNIMS, gas chromatography/electron capture negative ion mass spectrometry; GC/ECD, gas chromatography/electron capture detection; MS/MS, tandem mass spectrometry; ES, enzyme staining; FL, immunofluorescence microscopy.

[b] Estimated lower limit of detection is expected to reflect intrinsic sensitivity of test, i.e., estimated sensitivity in measuring same hypothetical genetic toxicant in each assay. Values are given only for comparison and can vary by several orders of magnitude, depending on physicochemical properties of the toxicant or adduct. 3,300 fmol/mg DNA corresponds to 1 DNA adduct per 10^6 nucleotides.

[c] DNA available from human specimens ranges from 300–500 μg in 40–50 ml of blood. This amount can be substantially greater if isolated from tissues obtained in surgery or at autopsy.

[d] Calculated assuming 0.5 mg of DNA is available.

[e] Roger W. Giese, Northeastern University, Boston, Mass.

[f] Paul Lohman, Sylvius Laboratory, Department of Radiation Genetics and Chemical Mutagenesis, State University of Leiden, Leiden, The Netherlands.

TABLE 2-2 Assays for Detection of Interaction Between Genetically Toxic Agents and Protein

Method for Detecting Protein Adducts[a]	Estimated Lower Limit[b] of Detection, adducts per unit of protein	Amount of Protein Needed for Assay[c]	Amount of Protein Available in Human Samples[c]	Advantages	Disadvantages
GC/HPLC	10 fmol/mg	50 mg blood	150 mg hemoglobin/ml blood	Large amount of material available, proteins accumulate more adducts than DNA	Biological significance unknown
Immunochemical	10 fmol/ml	5–500 ml urine	1,000 ml urine	Low costs, routine application possible	Biological significance doubtful, availability of antibodies
HPLC	10 fmol/ml	5–500 ml urine	1,000 ml urine	Low costs, routine application possible	Biological significance doubtful
GC/MS	1 fmol/ml	5–10 ml urine	1,000 ml urine	Multiple adducts detectable	Biological significance doubtful, sophisticated equipment needed

[a]Abbreviations: GC, gas chromatography; HPLC, high-performance liquid chromatography; MS, mass spectrometry.
[b]Estimated lower limit of detection is expected to reflect intrinsic sensitivity of test, i.e., estimated sensitivity in measuring same hypothetical genetic toxicant in each assay. Values are given only for comparison and can vary by several orders of magnitude, depending on physicochemical properties of toxicant or adduct.
[c]Estimates from Farmer et al., 1987.

opments in the study of DNA binding and protein binding provide a useful tool for beginning to acquire that understanding. However, additional information, such as clarification of the role of background or baseline adducts that are always present in animals and humans, will be needed to make full use of the advanced technology that is currently available.

APPLICATIONS OF DNA ADDUCT TECHNOLOGY

General Utility in Risk Assessment

DNA-adduct and protein-adduct technology is potentially useful in the processes of hazard identification and risk assessment. The NRC has performed comprehensive toxicological assessments for the Environmental Protection Agency on chemicals found in drinking water (NRC 1977, 1980a,b, 1982, 1983, 1986, 1987). In these assessments, information on acute, subchronic, and chronic effects is assembled and evaluated. Considerations of exposure and pharmacokinetics play an important role in risk assessment for chemicals with identified toxicity (NRC, 1987). It appears that adduct technology could be extremely valuable in estimating dosimetry and systemic distribution, in establishing possible target tissues or organs, and in determining the potential for irreversible toxicity such as cancer, mutation, or developmental effects.

Table 2-3 lists some potential applications of DNA-adduct analysis to the toxicologic evaluation of drinking water contaminants for risk assessment. The table is arranged as a matrix with the components of toxicity assessment on one axis and the potential contribution of DNA-adduct analysis on the other. The second column identifies when the specific method chosen to detect adducts is important. Toxic effects can be adduct-specific, and three toxicologic components—relationship of adducts to toxic response, mutagenicity, and species extrapolation—might require methods that identify the adducts detected. Some analytic methods (e.g., ^{32}P-postlabeling) do not identify the DNA adduct detected. The third column specifies whether or not it is desirable to identify the DNA adducts in using the technology for toxicologic assessments. Specific adduct identification is needed to correlate toxicity with adducts and is desirable in studies of mutagenic activity. Some DNA adducts, such as N7 alkylguanines, appear to have only a small role in mutation induction. Different components of toxicologic assessment have different requirements for quantitative or qualitative test results. These are identified in the table, as is the need for other biologic data. For example, information on chronicity of exposure is important in establishing a carcinogenicity hazard associated with a substance that induces formation of DNA adducts. The last column of the table identifies the extent to which DNA-adduct technology can now be used routinely in toxicologic assessments;

TABLE 2-3 Applications of DNA-Adduct Technology to Toxicologic Evaluations

Specific Components of Technological Assessments for Which DNA Adduct Technology Is Applicable	Is Application of DNA-Adduct Technology Method- or Technique-Specific?	Is DNA-Adduct Identification Desirable for This Application of DNA-Adduct Technology?	What Are the Minimal Results Needed for This Application of the Technology?	Are Other Biologic Data Needed for This Application?	How Routine Is DNA-Adduct Technology for This Application?
Exposure					
Qualitative	No	No	Qualitative[b]	No	Moderately
Dosimetry	No	No	Quantitative[c]	No	R&D[d]
Pharmacokinetic					
Tissue/organ distribution	No	No	Quantitative[b]	No	R&D
Relationship of adduct to toxicity	Yes	Yes	Quantitative	Yes	R&D
Hazard Assessment					
Cancer potential	No	Possibly[a]	Qualitative	Yes	Moderately
Mutagenic potential	Yes	Yes	Qualitative	Yes	Moderately
Genotoxicity	No	No	Qualitative	No	Moderately
Risk Assessment					
Extrapolation of dose	No	No	Quantitative	Yes	R&D
Species extrapolation	Possibly[a]	Possibly[a]	Qualitative	Yes	R&D
Target site extrapolation	No	No	Qualitative	Yes	R&D

[a] Adduct identification would increase ability to interpret significance of results.
[b] Number of adducts/10^6 nucleotide unnecessary.
[c] Number of adducts/10^6 nucleotide necessary.
[d] Studies are primarily at the research and development level.

most of the technology is still in the research and development stage, and none can yet be considered routine. The methods listed in the table are expected to evolve, so the conditions identified will also change. However, the table should facilitate an understanding of which aspects of toxicologic testing can be aided by DNA-adduct technology, which methods to consider first, and what data one should expect from a specific approach.

Features of the four categories of methods now available for use in toxicity testing are summarized in Table 2-4. Immunochemical and physical methods require considerable expertise in chemical synthesis, antibody production and radiolabeling, and analytic instrumentation. Some methods, such as ^{32}P-postlabeling, have been developed for use with high-molecular-weight (bulky) adducts, especially polycyclic aromatic hydrocarbons and aromatic amines. Use of this assay with low-molecularweight alkylating agents is not now feasible.

The published literature has been searched for specific information regarding the reactivity of 16 compounds that have been identified in drinking water and are known to be carcinogenic, mutagenic, teratogenic, or genetically toxic in experimental animals. Thirteen were recently reviewed by the NRC (1986) for EPA. Chemical structure, use, occurrence or source of exposure, association with DNA adduct formation, tissue distribution, carcinogenicity, mutagenicity, and other health effects were the data elements sought. The results are summarized in Appendix A, which can be referred to for some evidence for the theoretical assessments. Table 2-5 summarizes speculations about the ability of the 16 compounds to form DNA adducts. With the exception of benzo[a]pyrene, for which considerable data exists concerning adduct analysis, the chemicals have not been the subject of extensive DNA-binding studies.

Epidemiology and Human Monitoring

Proper investigation of relationships between disease in humans and exposure to drinking water contaminants has been hampered by the difficulty of assessing exposure to contaminants appropriately and by the limitations inherent in the use of traditional end points, such as the development of cancer, which are both rare and characterized by long latency. Interest in the incorporation of biologic markers into studies of human exposure to xenobiotic substances has increased, with the hope that use of such markers will enable scientists to characterize the empirical associations between exposures and outcomes, improve the accuracy of exposure assessment, enhance understanding of toxic mechanisms, increase the ability to detect early subclinical effects of exposure, and make better use of data from laboratory animals in predicting the effects of exposures of humans (NRC, 1987). Protein and DNA adducts in humans have been proposed as markers both

TABLE 2-4 Evaluation of New Molecular Methods in DNA- or Protein-Adduct Technology

| Characteristic | Method of Biologic Monitoring[a] for DNA Adducts | | | For Protein Adducts |
	Physical Methods	Immunochemical Methods	^{32}P-Post-Labeling	
Appropriateness for measuring exposure				
Qualitative	(+)	+	+	+
Recent (1-week) internal dose	?	+	+	+
Long-term body burden	?	+	+	(+)
Dose at target site	?	+	+	−
Appropriateness for assessing health effects				
Reversible	?	(−)	(−)	(−)
Irreversible	(−)	(−)	(−)	(−)
Interpretation of results				
On individual basis	+	+	+	+
On group basis	+	+	+	+
Precision of method				
Technical reproducibility	?	(+)	(+)	+
Stability	(+)	(+)	+	(+)
Interlaboratory reproducibility	?	(+)	(+)	+
Sensitivity				
For some environmental exposures	?	+	+	+
For occupational exposures	(+)	+	+	+
For acute exposures	+	+	+	+
Chemical specificity	+	[b]	−	[b]
Absence of confounding factors	?	(+)	(+)	(+)
Absence of background adducts	?	(−)	(−)	(−)
Simplicity	−	[b]	[b]	[b]
Ease of sample storage	+	+	+	+
Current applicability				
In research	(+)	+	+	+
In routine use	(−)	(+)	(+)	(+)

[a]Symbols: +, applicable or true; (+), probably applicable or true; −, not applicable or not true; (−), not now applicable or not now true; ?, unknown.
[b]Cannot be generalized.

TABLE 2-5   Classification of 16 Drinking Water Contaminants According to Their Presumed Ability to Form DNA Adducts

Definite Ability	Probable Ability	Possible Ability	Insufficient Data
Acrylamide	Trichlorfon	Diallate	Arsensic
Chromium		Sulfallate	Nitrofen
Benzo[a]pyrene		1,2-Dichloropropane	Pentachlorophenol
Dibromochloro-		(1,2-DCP)	
propane (DBCP)		1,2,3-Trichloro-	
Ethylene di-		propane (1,2,3-TCP)	
bromide (EDB)		1,3-Dichloropropene	
		(1,3-DCP)	
		Di(2-ethylhexyl)	
		phthalate (DEHP)	
		Mono(2-ethylhexyl)	
		phthalate (MEHP)	

for use in epidemiologic studies to assess the risks associated with exposure to potential genetic toxicants and for use in monitoring exposed populations. However, all the studies that have measured protein or DNA adducts have focused on humans exposed to carcinogens—occupationally, environmentally, or otherwise.

In principle, incorporation of measurements of carcinogen-DNA adduct formation into epidemiologic studies could offer at least two kinds of benefits:

• The use of sensitive methods, such as immunoassays and ^{32}P-postlabeling, might afford an opportunity to detect early, subtle effects of small exposures.

• Human studies incorporating DNA-adduct assays might provide information on target-molecule dose that reflects exposure, absorption, metabolism, and DNA-adduct formation and repair rates.

Although the measurement of DNA-adduct formation in humans holds substantial promise for epidemiologic and monitoring studies, interpretation of data derived from DNA-adduct measurements is extremely complex, particularly in humans. In general, further experimental work is required before measurements of DNA adducts can be successfully incorporated into studies that assess toxicity from drinking water contaminants in humans. The following issues are of particular concern in considering the potential applications of DNA-adduct technology in human studies to evaluate potential toxicity of drinking water contaminants:

• The paucity of information about the kinetics, dose-response relationships, and interindividual and intraindividual variability of DNA-adduct formation in humans renders the proper design and interpretation of human studies using DNA-adduct technology very difficult. Studies that characterize the variability of adduct levels in humans due to such factors as age, sex, ethnicity, diet, tobacco use, and such medical conditions as liver disease are needed. Additionally, studies are needed for proper characterization of baseline adduct levels in the general population.

• Because numbers of DNA adducts reflect not only exposure, but also rates of metabolism (in the case of indirect carcinogens) and DNA-adduct formation and repair, DNA-adduct concentrations are likely to involve complex dynamics. When technically feasible, the use of protein adducts might prove more appropriate for exposure assessment; such adducts in the hemoglobin of red blood cells have demonstrated chemical stability and linear dose-response relationships for a variety of compounds and thus can provide integrated exposure information. To date, the use of protamine adducts in germ cells has been limited to studies of small alkylating agents in mice. Further studies are needed to evaluate the use of human protamines in dosimetry. Protamine dosimetry might help to identify the exposures that pose germinal risks.

• In animal studies, it is possible to study DNA-adduct concentrations in target tissue, but the target tissue of interest in humans is often inaccessible. Circulating white blood cells or lymphocytes are used as surrogates for determination of DNA-adduct concentrations. However, the validity of using surrogate tissue, particularly for human risk assessment, has not been adequately evaluated.

• Some of the better-characterized chemicals that produce DNA adducts, such as BaP, are ubiquitous in the environment. That presents difficulties in epidemiologic studies, because, even with proper selection of controls, background DNA-adduct concentrations might mask slight differences in concentrations between "exposed" and "unexposed" populations.

REFERENCES

Adamkiewicz, J., G. Eberle, N. Huh, P. Nehls, and M. F. Rajewsky. 1985. Quantitation and visualization of alkyl deoxynucleosides in the DNA of mammalian cells by monoclonal antibodies. Environ. Health Perspect. 62:49–55.

Adams, J., M. David, and R. W. Giese. 1986. Pentafluorobenzylation of O^4-ethylthymidine and analogs by phase-transfer catalysis for determination by gas chromatography with electron capture detection. Anal. Chem. 58:345–348.

Adriaenssens, P. I., C. M. White, and M. W. Anderson. 1983. Dose-response relationships for the binding of benzo(a)pyrene metabolites to DNA and protein in lung, liver, and forestomach of control and butylated hydroxyanisole-treated mice. Cancer Res. 43:3712–3719.

Baan, R. A., P. H. M. Lohman, A. M. J. Fichtinger-Schepman, M. A. Muysken-Schoen, and J. S. Ploem. 1986. Immunochemical approach to detection and quantitation of DNA adducts resulting from exposure to genotoxic agents. Prog. Clin. Biol. Res. 207:135–146.

Baan, R. A., P. T. M. van den Berg, M.-J. S. T. Steenwinkel, and C. J. M. van der Wulp. 1988. Detection of benzo[*a*]pyrene-DNA adducts in cultured cells treated with benzo[*a*]pyrene diol epoxide by quantitative immunofluorescence microscopy and ^{32}P-postlabelling; immunofluorenscence analysis of benzo[*a*]pyrene-DNA adducts in bronchial cells from smoking individuals. Pp. 146–156 in Methods for Detecting DNA Damaging Agents in Humans: Applications in Cancer Epidemiology and Prevention, H. Bartsch, K. Hemminki, and I. K. O'Neill, eds. IARC Scientific Publications, No. 89. Lyon: International Agency for Research on Cancer.

Bailey, E., P. B. Farmer, and D. E. Shuker. 1987. Estimation of exposure to alkylating carcinogens by the GC-MS determination of adducts to hemoglobin and nucleic acid bases in urine. Arch. Toxicol. 60:187–191.

Bartsch, H., K. Hemminki, and I. K. O'Neill, eds. 1988. Methods for Detecting DNA Damaging Agents in Humans: Applications in Cancer Epidemiology and Prevention. IARC Scientific Publications, No. 89. Lyon: International Agency for Research on Cancer. 5 pp.

Berlin, A., M. Draper, K. Hemminki, and H. Vaino, eds. 1984. Monitoring Human Exposure to Carcinogenic and Mutagenic Agents. IARC Scientific Publications No. 59. Lyon: International Agency for Research on Cancer. 457 pp.

de Serres, F. J. 1988. Banbury Center DNA Adducts workshop. Workshop on DNA Adducts, Banbury Center, Cold Spring Harbor Laboratory, Cold Spring Harbor, New York, NY (U.S.A.), September 30–October 2, 1986. Mutat. Res. 203:55–68.

de Serres, F. J., B. L. Gledhill, and W. Sheridan, eds. 1985. DNA adducts: Dosimeters to monitor human exposure to environmental mutagens and carcinogens, September 24–26, 1984, Research Triangle Park, North Carolina. Environ. Health Perspect. 62:1–238.

Den Engelse, L., P. M. A. B. Terhaggen, and A. C. Begg. 1988. Cisplatin–DNA interaction products in sensitive and resistant cell lines, and in buccal cells from cisplatin-treated cancer patients. Proc. Am. Assoc. Cancer Res. 29:339 (Abstract 1348).

Ehrenberg, L. 1988. Dose monitoring and cancer risk. Pp. 23–31 in Methods for Detecting DNA Damaging Agents in Humans: Applications in Cancer Epidemiology and Prevention, H. Bartsch, K. Hemminki, and I. K. O'Neill, eds. IARC Scientific Publications, No. 89. Lyon: International Agency for Research on Cancer.

Farmer, P. B., H.-G. Neumann, and D. Henschler. 1987. Estimation of exposure of man to substances reacting covalently with macromolecules. Arch. Toxicol. 60:251–260.

Farmer, P. B., J. Lamb, and P. D. Lawley. 1988. Novel uses of mass spectrometry in studies of adducts of alkylating agents with nucleic acids and proteins. Pp. 347–355 in Methods for Detecting DNA Damaging Agents in Humans: Applications in Cancer Epidemiology and Prevention, H. Bartsch, K. Hemminki, and I. K. O'Neill, eds. IARC Scientific Publications, No. 89. Lyon: International Agency for Research on Cancer.

Giese, R. W. 1986. Determination of DNA adducts by electropore labeling-GC. Pp. 41–55 in Environmental Epidemiology, F. C. Koufler and G. F. Craun, eds. Chelsea, Mich.: Louis Publishers.

Gupta, R. C., and K. Randerath. 1988. Analysis of DNA adducts by ^{32}P-postlabeling and thin-layer chromatography. Pp. 401–420 in DNA Repair: A Laboratory Manual of Research Procedures, Vol. 3, E. C. Friedberg and P. C. Hanawalt, eds. New York: Marcel Dekker.

Gupta, R. C., M. V. Reddy, and K. Randerath. 1982. ^{32}P-postlabeling analysis of non-radioactive aromatic carcinogen–DNA adducts. Carcinogenesis 3:1081–1092.

Harris, C. C., R. H. Yolken, and I.-C. Hsu. 1982. Enzyme immunoassays: Applications in cancer research. Methods Cancer Res. 20:213–243.

Heisig, V., A. M. Jeffrey, M. J. McGlade, and G. J. Small. 1984. Fluorescence-linenarrowed spectra of polycyclic aromatic carcinogen-DNA adducts. Science 223:289–291.

Huitfeldt, H. S., E. F. Spangler, J. M. Hunt, and M. C. Poirier. 1986. Immunohistochemical localization of DNA adducts in rat liver tissue and phenotypically altered foci during oral administration of 2-acetylaminofluorene. Carcinogenesis 7:123–129.

Huitfeldt, H. S., E. F. Spangler, J. Baron, and M. C. Poirier. 1987. Microfluorometric determination of DNA adducts in immunofluorescent-stained liver tissue from rats fed 2-acetylaminofluorene. Cancer Res. 47:2098–2102.

Jankowiak, R., R. S. Cooper, D. Zamzow, G. J. Small, G. Doskocil, and A. M. Jeffrey. 1988. Enhancement of sensitivity of fluorescence line narrowing spectrometry for detection of carcinogen–DNA adducts. Pp. 372–377 in Methods for Detecting DNA Damaging Agents in Humans: Applications in Cancer Epidemiology and Prevention, H. Bartsch, K. Hemminki, and I. K. O'Neill, eds. IARC Scientific Publications, No. 89. Lyon: International Agency for Research on Cancer.

Knox, R. J., F. Friedlos, D. A. Lydall, and J. J. Roberts. 1986. Mechanism of cytotoxicity of anticancer platinum drugs: Evidence that cis-diamminedichloroplatinum(II) and cis-diammine-(1,1-cyclobutanedicarboxylato)platinum(II) differ only in the kinetics of their interaction with DNA. Cancer Res. 46:1972–1979.

Lohman, P. H. M., J. D. Jansen, and R. A. Baan. 1984. Comparison of various methodologies with respect to specificity and sensitivity in biomonitoring occupational exposure to mutagens and carcinogens. Pp. 259–277 in Monitoring Human Exposure to Carcinogenic and Mutagenic Agents, A. Berlin, M. Draper, K. Hemminki, and H. Vainio, eds. IARC Scientific Publications, No. 59. Lyon: International Agency for Research on Cancer.

Lohman, P. H. M., R. A. Baan, A. M. J. Fichtinger-Schepman, M. A. Muysken-Schoen, R. J. Lansbergen, and F. Berends. 1985. Molecular dosimetry of genotoxic damage: Biochemical and immunochemical methods to detect DNA-damage. Trends Pharmacol. Sci. (Nov. FEST Suppl.):1–7.

Minnetian, O., M. Saha, and R. W. Giese. 1987. Oxidation-elimination of a DNA base from its nucleoside to facilitate determination of alkyl chemical damage to DNA by gas chromatography with electrophore detection. J. Chromatogr. 410:453–457.

Mohamed, G. B., A. Nazareth, M. J. Hayes, R. W. Giese, and P. Vouros. 1984. Gas chromatography-mass spectrometry characteristics of methylate perfluoroacyl derivatives of cytosine and 5-methylcytosine. J. Chromatogr. 314:211–217.

Müller, R., and M. F. Rajewsky. 1981. Antibodies specific for DNA components structurally modified by chemical carcinogens. J. Cancer Res. Clin. Oncol. 102:99–113.

Neumann, H.-G. 1984. Review: Analysis of hemoglobin as a dose monitor for alkylating and arylating agents. Arch. Toxicol. 56:1–6.

Neumann, H.-G. 1988. Haemoglobin binding in control of exposure to and risk assessment of aromatic amines. Pp. 157–165 in Methods for Detecting DNA Damaging Agents in Humans: Applications in Cancer Epidemiology and Prevention, H. Bartsch, K. Hemminki, and I. K. O'Neill, eds. IARC Scientific Publications, No. 89. Lyon: International Agency for Research on Cancer.

NRC (National Research Council). 1977. Drinking Water and Health, Vol. 1. Washington, D.C.: National Academy Press. 939 pp.

NRC (National Research Council). 1980a. Drinking Water and Health, Vol. 2. Washington, D.C.: National Academy Press. 393 pp.

NRC (National Research Council). 1980b. Drinking Water and Health, Vol. 3. Washington, D.C.: National Academy Press. 415 pp.

NRC (National Research Council). 1982. Drinking Water and Health, Vol. 4. Washington, D.C.: National Academy Press. 299 pp.

NRC (National Research Council). 1983. Drinking Water and Health, Vol. 5. Washington, D.C.: National Academy Press. 157 pp.

NRC (National Research Council). 1986. Drinking Water and Health, Vol. 6. Washington, D.C.: National Academy Press. 457 pp.

NRC (National Research Council). 1987. Drinking Water and Health, Disinfectants and Disinfectant By-Products, Vol. 7. Washington, D.C.: National Academy Press. 207 pp.

NRC (National Research Council), Committee on Biological Markers. 1987. Biological markers in environmental health research. Environ. Health Perspect. 74:3–9.

Ohshima, H., and H. Bartsch. 1988. Urinary N-nitrosamino acids as an index of exposure to N-nitroso compounds. Pp. 83–91 in Methods for Detecting DNA Damaging Agents in Humans: Applications in Cancer Epidemiology and Prevention, H. Bartsch, K. Hemminki, and I. K. O'Neill, eds. IARC Scientific Publications, No. 89. Lyon: International Agency for Research on Cancer.

Osterman-Golkar, S. 1988. Dosimetry of ethylene oxide. Pp. 249–257 in Methods for Detecting DNA Damaging Agents in Humans: Applications in Cancer Epidemiology and Prevention, H. Bartsch, K. Hemminki, and I. K. O'Neill, eds. IARC Scientific Publications, No. 89. Lyon: International Agency for Research on Cancer.

Pereira, M. A., L.-H. C. Lin, and C. W. Chang. 1981. Dose-dependency of 2-acetylaminofluorene binding to liver DNA and hemoglobin in mice and rats. Toxicol. Appl. Pharmacol. 60:472–478.

Perera, F. P., R. M. Santella, D. Brenner, T.-L. Young, and I. B. Weinstein. 1988. Application of biological markers to the study of lung cancer causation and prevention. Pp. 451–459 in Methods for Detecting DNA Damaging Agents in Humans: Applications in Cancer Epidemiology and Prevention, H. Bartsch, K. Hemminki, and I. K. O'Neill, eds. IARC Scientific Publications, No. 89. Lyon: International Agency for Research on Cancer.

Poirier, M. C. 1981. Antibodies to carcinogen–DNA adducts. J. Natl. Cancer Inst. 67:515–519.

Poirier, M. C. 1984. The use of carcinogen–DNA adduct antisera for quantitation and localization of genomic damage in animal models and the human population. Environ. Mutag. 6:879–887.

Rahn, R. O., S. S. Chang, J. M. Holland, and L. R. Shugart. 1982. A fluorometric-HPLC assay for quantitating the binding of benzo(a)pyrene metabolites to DNA. Biochem. Biophys. Res. Commun. 109:262–268.

Randerath, K., R. H. Miller, D. Mittal, and E. Randerath. 1988. Monitoring human exposure to carcinogens by ultrasensitive postlabelling assays: Application to unidentified genotoxicants. Pp. 361–367 in Methods for Detecting DNA Damaging Agents in Humans: Applications in Cancer Epidemiology and Prevention, H. Bartsch, K. Hemminki, and I. K. O'Neill, eds. IARC Scientific Publications, No. 89. Lyon: International Agency for Research on Cancer.

Sanders, M. J., R. S. Cooper, R. Jankowiak, G. J. Small, V. Heisig, and A. M. Jeffrey. 1986. Identification of polycyclic aromatic hydrocarbon metabolites and DNA adducts in mixtures using fluorescence line narrowing spectrometry. Anal. Chem. 58:816–820.

Santella, R. M. 1988. Application of new techniques for the detection of carcinogen adducts to human population monitoring. Mutat. Res. 205:271–282.

Santella, R. M., F. Gasparo, and L. L. Hsieh. 1987. Quantitation of carcinogen–DNA adducts with monoclonal antibodies. Prog. Exp. Tumor Res. 31:63–75.

Shugart, L. 1985. Quantitating exposure to chemical carcinogens: In vivo alkylation of hemoglobin by benzo[a]pyrene. Toxicology 34:211–220.

Shuker, D. E. G., and P. B. Farmer. 1988. Urinary excretion of 3-methyladenine in humans as a marker of nucleic acid methylation. Pp. 92–96 in Methods for Detecting DNA Damaging Agents in Humans: Applications in Cancer Epidemiology and Prevention, H. Bartsch, K.

Hemminki, and I. K. O'Neill, eds. IARC Scientific Publications, No. 89. Lyon: International Agency for Research on Cancer.

Umbenhauer, D., C. P. Wild, R. Montesano, R. Saffhill, J. M. Boyle, N. Huh, U. Kirsten, J. Thomale, M. F. Rajewsky, and S. H. Lu. 1985. O^6-methyldeoxyguanosine in oesophageal DNA among individuals at high risk of oesophageal cancer. Int. J. Cancer 36:661–665.

Van Benthem, J., C. P. Wild, E. Vermeulen, H. H. K. Winterwerp, L. Den Engelse, and E. Scherer. 1988. Immunocytochemical localization of DNA adducts in rat tissues following treatment with N-nitrosomethylbenzylamine. Pp. 102–106 in Methods for Detecting DNA Damaging Agents in Humans: Applications in Cancer Epidemiology and Prevention, H. Bartsch, K. Hemminki, and I. K. O'Neill, eds. IARC Scientific Publications, No. 89. Lyon: International Agency for Research on Cancer.

Vanderlaan, M., B. E. Watkins, M. Hwang, M. G. Knize, and J. S. Felton. 1988. Monoclonal antibodies for the immunoassay of mutagenic compounds produced by cooking beef. Carcinogenesis 9:153–160.

van Sittert, N. J. 1984. Biomonitoring of chemicals and their metabolites. Pp. 153–172 in Monitoring Human Exposure to Carcinogenic and Mutagenic Agents, A. Berlin, M. Draper, K. Hemminki, and H. Vainio, eds. IARC Scientific Publications, No. 59. Lyon: International Agency for Research on Cancer.

Wild, C. P., R. C. Garner, R. Montesano, and F. Tursi. 1986. Aflatoxin B_1 binding to plasma albumin and liver DNA upon chronic administration to rats. Carcinogenesis 7:853–858.

Wogan, G. 1988. Summary. Pp. 9–12 in Methods for Detecting DNA Damaging Agents in Humans: Applications in Cancer Epidemiology and Prevention, H. Bartsch, K. Hemminki, and I. K. O'Neill, eds. IARC Scientific Publications, No. 89. Lyon: International Agency for Research on Cancer.

3

Conclusions and Recommendations

The subcommittee has reached the following conclusions regarding the application of DNA-adduct and protein-adduct assays to EPA's assessment of drinking water contaminants:

- Tests that detect and measure DNA adducts and protein adducts can be used as quantitative and qualitative dosimeters of exposure and may permit calibration of internal versus external dose across several orders of magnitude for specific genetic toxicants.
- The presence of DNA adducts or protein adducts might not in itself establish a risk, and tests for DNA adducts or protein adducts should not be used in isolation for hazard or risk assessments.
- Current evidence suggests an association between the onset of specific types of toxicity (mutation, cancer, or developmental effects) and the concentration of DNA adducts. The toxic effect is usually tissue-specific, although adducts can form in many tissues. The subcommittee cannot define a general concentration of adducts that might be tolerated by animals, but an evaluation in Chapter 1 of the limited in vitro and in vivo data indicates that concentrations of 1–10,000 adducts per 10^7 normal bases are associated with adverse effects. Most of the methods described in Tables 2-1 and 2-2 can span that range.
- Most current models for risk assessment do not incorporate such data as DNA alterations over a broad range of exposures. Risk assessments might be improved if new kinds of mathematical models can incorporate biologic data from studies of DNA damage over a range of doses to show where detoxification and DNA repair occur and reach their limits. However, such

models will need to be carefully evaluated before they are put into general use.

• DNA-adduct detection methods, especially the ^{32}P-postlabeling method, have demonstrated the presence of considerable persistent (mostly unknown) DNA lesions in various target cells of untreated animals. The toxicologic importance of those "background" lesions is unknown; they might reflect endogenous exposures from normal body constituents or processes or from naturally occurring mutagens and carcinogens in the environment, such as ultraviolet radiation and mutagens and carcinogens in foods and water.

• If the goal is to assess exposure to a drinking water contaminant identified as genetically toxic, it might sometimes be more appropriate to use protein adducts as dosimeters. Although their biologic significance could be quite different from that of DNA adducts, hemoglobin adducts have demonstrated chemical stability and linear dose-response relationships for a variety of compounds. Germ cell studies have suggested that protamine adducts are relevant to genetic risk assessment.

• Monitoring of the formation of DNA adducts or protein adducts among exposed humans, and among appropriate controls, holds promise for use in assessing the human risk associated with consumption of drinking water contaminated by potentially genotoxic agents. However, available information on baseline DNA-adduct concentrations, adduct persistence, interindividual and intraindividual variability, dose-response relationships, and biologic significance in humans is scanty. In addition, human tissues available for such studies may not be the tissues relevant for the major toxic end point, such as carcinogenesis. Interpretation of epidemiologic or monitoring studies of humans would therefore be difficult.

The subcommittee recognizes that full application of adduct technology to toxicologic assessments would require the filling of gaps in information and advances in technology. The subcommittee offers the following recommendations to EPA:

• Investigations into the origin and role of the natural background of DNA adducts are needed. The resulting information will be relevant to the interpretation of DNA alterations induced by occupation, lifestyle, or environmental exposure.

• Because most of the data describing the induction of DNA alterations are derived from acute-exposure studies, investigations aimed at understanding the induction and persistence of DNA alterations over a broad range of doses in subchronic and chronic regimens should be conducted and correlated with other toxic end points.

• Investigations are needed to improve understanding of the kinetics and relative significance of DNA alterations and protein alterations as measures of somatic and germ cell risk in animals.

• Animal models should be used to study the use of peripheral white blood cells as target cells for in vivo analyses of DNA adducts, because such cells would be most accessible in human studies.

• Baseline data should be established on chemicals in drinking water that are presumed to be genetically toxic, with an eye to revealing qualitative and quantative associations between DNA or protein alterations and other components of hazard identification.

• Methods for detection, measurement, and identification of DNA adducts and protein adducts should be developed and validated, particularly for low-molecular-weight monofunctional and bifunctional aromatic and hydrophobic alkylating agents. Many drinking water contaminants (acrylamide, EDB, DBCP, etc.) are of this size and type.

• Differences between in vivo DNA-adduct formation and repair in somatic and germ cells should be studied. Although risk assessment using DNA-adduct measurements usually focuses on tumor formation, heritable effects of genetic toxicants should be considered a major burden for the human population.

A repository of tissue samples with known DNA-adduct or protein-adduct concentrations should be established, from which a calibration process might be developed.

APPENDIX **A**

Drinking Water Contaminant Candidates for Development of Baseline Data on Formation of DNA Adducts

This appendix presents data on 16 compounds found in drinking water—13 are contaminants recently reviewed by the NRC (1986) for EPA—that are known to be teratogenic, carcinogenic, mutagenic, or genetically toxic in laboratory animals. Of these, five—acrylamide, benzo[a]pyrene (BaP), chromium, dibromochloropropane (DBCP), and ethylene dibromide (EDB)—have definitely been shown to form DNA adducts. Trichlorfon probably forms DNA adducts. On the basis of current evidence, diallate, sulfallate, the two chloropropanes, and chloropropene can be considered possible inducers of DNA adducts.

Evidence on the five remaining compounds—arsenic, nitrofen, pentachlorophenol, and the ethylhexyl phthalates—is insufficient to permit classifying them as inducers of DNA adducts. They are included because they have been shown to be mutagens or carcinogens or because they produce genetic damage of some kind. They might form DNA adducts that have not yet been detected.

CONTAMINANTS THAT DEFINITELY FORM DNA ADDUCTS

Acrylamide

$$CH_2 = \overset{\overset{\displaystyle H}{\displaystyle |}}{C} - \overset{\overset{\displaystyle O}{\displaystyle \|}}{C} - NH_2$$

Acrylamide is a monomer of polyacrylamide. It is highly reactive and

61

reacts spontaneously with hydroxyl-, amino-, and sulfhydryl-containing compounds. With a solubility in water of 219 g/100 ml, acrylamide polymers are used to improve oil recovery, increase dry strength of paper products, dissipate fog, and stabilize soil. Acrylamides are also used in grouting operations, clarification of potable water, and treatment of municipal and industrial effluents. Acrylamide is biodegradable. It is degraded to carbon dioxide in 4–12 days in water and is completely degraded in 6 days in soil. Up to 60% is degraded to carbon dioxide (NRC, 1986).

Occurrence   No estimate of occurrence in drinking water has been calculated.

Tissue Distribution   Acrylamide is rapidly metabolized in the body, 24 hours after oral administration of acrylamide no trace of the parent compound was found in any tissues. Metabolites of acrylamide accumulate in erythrocytes. The half-life of the metabolites in erythrocytes is 10.5 days, and in other tissues 8 days (as measured by the decrease in radioactivity using ^{14}C-labelled acrylamide) (Miller et al., 1982).

Formation of DNA Adducts   Solomon et al. (1985) studied the in vitro reaction of acrylamide with 2′-deoxynucleosides and calf thymus DNA. Acrylamide reacted most strongly at the N6 position of 2′-deoxyadenosine and, to a lesser extent, at the N3 position of 2′-deoxycytidine, although adduct formation of other 2′-deoxynucleosides was recorded as well. With calf thymus DNA, acrylamide reacted strongest at the N1 position of 2-deoxyadenosine. The authors point out that adducts formed were only present in small quantities, even after the long reaction time (10 or 40 days), so the results cannot be readily translated to human exposure to acrylamide. However, the possibility of predicting in vivo alkylation with direct-acting alkylating agents on the basis of in vitro adduct formation has been suggested.

Mutagenicity   Mutagenicity studies with acrylamide in the *Salmonella/* microsome assay were negative. Acrylamide is clastogenic in L5178Y mouse lymphoma cells (Moore et al., 1987). It is known to induce dominant lethal mutations in male rodents, and induced translocations in postmeiotic germ cells of mice in a recent study (Shelby et al., 1987).

Carcinogenicity   A number of reports have shown acrylamide to be carcinogenic in laboratory animals, although no data have shown that for humans. Male and female Fischer 344-rats have developed tumors after a 2-year exposure to acrylamide in drinking water. Male rats exposed to acrylamide at 0.5 mg/kg of body weight each day for 2 years developed scrotal mesotheliomas. At 2.0 mg/kg per day, benign thyroid tumors, malignant thyroid tumors, glial tumors in the CNS, adenomas of the clitoral gland, squamous cell papillomas in the mouth, benign and malignant mammary tumors, and malignant uterine tumors were observed in rats. On the basis of a multistage model, and assuming the consumption of 1 liter of water per day containing acrylamide at 1 μg/liter, the human lifetime carcinogenic risk is estimated at $3.8–8.2 \times 10^{-6}$ and the upper 95% confidence estimate of lifetime cancer risk is $0.75–1.4 \times 10^{-5}$ (NRC, 1986).

Other Health Effects   Acrylamide has been reported to cause neuropathy in humans that leads to progressive symmetric distal sensory abnormalities and motor weakness. Also stated to occur were slurred speech; unsteady gait; memory loss; irrational behavior; visual, tactile, and auditory hallucinations; skin sensitization; cold blue hands; muscle weakness; paresthesia; and numbness of hands or feet. Neuropathy has been reported in animals (cats, rats, mice, guinea pigs, rabbits, and monkeys) that were given acrylamide (NRC, 1986).

Summary   Acrylamide can produce peripheral neuropathy in animals and humans. It is a known carcinogen in animals.

Benzo[*a*]pyrene (B*a*P)

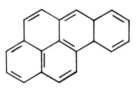

B*a*P is a ubiquitous polycyclic aromatic hydrocarbon associated with combustion. It was first isolated from coal tar.

Occurrence   B*a*P concentrations of 3 ng/liter in tap water were reported in a study of European water distribution systems. Concentrations of less than 1 ng/liter were found in a study of six U.S. drinking water systems (NRC, 1982).

Tissue Distribution   B*a*P has been found in the liver, lungs, colon, kidneys, muscle, brain, and forestomach of mice. It tends to localize in fatty tissues.

Formation of DNA Adducts   Cytochrome P-450 enzymes and epoxide hydrolases metabolize B*a*P to the two diastercoisomers responsible for DNA-adduct formation. The two diastercoisomers are (+)-7b,8a-dihydroxy-9a,10a-epoxy-7,8,9,10-tetrahydrobenzo[a]pyrene (BPDE I) and (−)-7b,8a-dihydroxy-9b,10b-epoxy-7,8,9,10-tetrahydrobenzo[a]pyrene (BPDE II). These two compounds bind mainly to 2-amino groups of guanine residues, but also to the N7 position (NRC, 1982; Sontag, 1981). BaP also forms N^6-adenine adducts (Jeffrey et al., 1979), which appear to be important in *ras* oncogene activation.

In rats, an increase in intravenous administration of BaP from 1 μmol to 10 μmol increased adduct concentration in rat lung approximately fivefold and in rat liver about threefold. The pattern of metabolism of BaP was similar in human, monkey, dog, hamster, and rat, but adduct concentration varied widely. The concentration was 30 times higher in human bladder than in rat bladder. In the trachea and bronchi, the incidence of adduct formation differed by a factor of 10 between human and rat.

Mutagenicity BaP is a positive mutagen in the *Salmonella*/microsome test (McCann et al., 1975).

Carcinogenicity Carcinogenic effects were observed in mice after the administration of 40–45 ppm BaP orally for 110 days (Rigdon and Neal, 1966, 1969). Rats and hamsters have also shown sensitivity to the induction of benign and malignant skin tumors by BaP (NRC, 1977). It is suspected that a possible consequence of long-term exposure to BaP is an increased incidence of bronchial carcinoma.

Other Health Effects Repeated oral BaP administration in mice resulted in hypoplastic anemia. Its acute toxicity is low.

Summary BaP is mutagenic and carcinogenic in animals and a suspected human carcinogen.

Chromium

Cr

Chromium is usually found as either hexavalent chromium, Cr(VI), or trivalent chromium, Cr(III). Cr(III) is an essential micronutrient that is low in the typical American diet. It is toxic at high doses. Relatively large quantities of chromium have been found in wastewater from the plating and finishing industries. Cr(VI) is much more toxic than Cr(III).

Occurrence A study of more than 1,500 surface waters in the United

States showed a maximal chromium content of 0.11 mg/liter and a mean of 0.01 mg/liter (NRC, 1980).

Tissue Distribution   In humans, chromium occurs at the highest concentration in the lungs. That suggests that the primary human exposure is through air, and not through food or water. Cr(III) is inhibited by membrane barriers; Cr(VI) is absorbed and then reduced to Cr(III) (NRC, 1977). Chromium concentrations of 1–5 µg/ml of plasma and less than 1 ng/ml of urine have been reported (NRC, 1980).

Formation of DNA Adducts   Cr(III) binds to free nucleotides and to nucleic acids at nucleophilic sites (such as the unesterified oxygen of phosphate groups and the nitrogen and oxygen atoms of nitrogen bases). Binding to DNA involves regions with high concentrations of guanine and cytosine and produces DNA–DNA cross-links that, if not repaired, lead to extensive DNA fragmentation. Cr(VI) causes breakage of polynucleotide chains that is inferred by changes in physicochemical properties of DNA (Merian et al., 1985).

Mutagenicity   Chromium decreases the fidelity of DNA synthesis and induces gene mutations, chromosomal aberrations, sister-chromatid exchanges, and malignant transformation in mammalian cells (Merian et al., 1985).

Carcinogenicity   Cr(VI) has been shown to cause dermal ulcerations and respiratory cancer. Cr(VI) is considered carcinogenic; the carcinogenicity of Cr(III) has not been conclusively established (NRC, 1986).

Other Health Effects   In humans, allergic contact dermatitis may result from exposure to either Cr(VI) or Cr(III). Workers in chromate plants have reported respiratory injury, most commmonly manifested as ulceration and perforation of the nasal septum (U.S. EPA, 1984).

Summary   Both Cr(VI) and Cr(III) have exhibited a potential for genetic toxicity. However, the potential of Cr(III) for genetic toxicity is much less than that of Cr(VI), due to the inability of Cr(III) to cross cellular membranes.

Dibromochloropropane (DBCP)

$$
\begin{array}{ccc}
H & H & H \\
| & | & | \\
H-C-C-C-H \\
| & | & | \\
Br & Br & Cl
\end{array}
$$

DBCP is a short-chain aliphatic halogenated hydrocarbon that is used as a soil fumigant and nematocide.

Occurrence   No data on DBCP concentrations in drinking water were found.

Tissue Distribution   After exposure, DBCP is found mainly in the kidneys and testes.

Formation of DNA Adducts   DBCP induces sister-chromatid exchanges and chromosomal aberrations. It also causes unscheduled DNA synthesis and thus can be considered an adduct-forming agent, but the mechanism is unknown. Radiolabeling data indicate that DBCP, or a metabolite, binds to DNA (Kato et al., 1980).

Mutagenicity   DBCP is an indirect mutagen in bacteria, in that it requires metabolic activation to induce mutagenic effects. In *Drosophila melanogaster*, DBCP caused the loss of X and Y chromosomes and induced increases in heritable translocations. DBCP induced sister-chromatid exchanges and chromosomal aberrations in Chinese hamster ovary cells. In humans, it caused an increase in the number of sperm containing two Y chromosomes, suggesting irreversible genetic change. DBCP induced unscheduled DNA synthesis (DNA repair) in premeiotic germ cells in prepubertal mice (NRC, 1986). Dominant lethal mutations were induced in male rats, but not in male mice, in a study by Teramoto et al. (1980). Sasaki et al. (1986) showed that DBCP is mutagenic for somatic cells of mice in vivo.

Carcinogenicity   DBCP induced a high incidence of squamous cell carcinoma of the forestomach and toxic nephropathy in male and female rats and mice and a high incidence of mammary adenocarcinoma in female rats. An upper 95% confidence estimate of lifetime risk for human males drinking 1 liter of water per day containing DBCP at 1 μg/liter is 9.9×10^{-6} (NRC, 1986).

Other Health Effects   Human exposure to DBCP has been shown to result in azoospermia or severe oligospermia. Reduction in sperm production is accompanied by increases in serum concentrations of follicle-stimulating hormone and luteinizing hormone and by reduction in or absence of spermatogenic cells in seminiferous tubules. The reduction in spermatogenesis might also involve disturbances in genetic material in the sperm, such as the increased frequency of Y-chromosome nondisjunction found in 18 DBCP-exposed workers by Kapp et al. (1979).

Male rats given DBCP at 120 mg/kg of body weight showed marked cytoplasmic vacuolization of the proximal renal tubular epithelium in the outer medulla and hepatic centrilobular necrosis. DBCP at 40 mg/kg produced only mild hepatocellular swelling in the periportal region of the liver lobules.

The testes and epididymides of the rats revealed relatively minor cellular injury 24 hours after the two highest doses (80 and 120 mg/kg) (NRC, 1986).

Summary   DBCP produces functional disturbances of the liver and kidneys and is carcinogenic in several organ systems in rats and mice. It is mutagenic in rats, and has adverse effects on human male fertility.

Ethylene Dibromide (EDB)

EDB has been used as an antiknock constituent in gasoline containing tetraethyl lead, as a fumigant-insecticide, as a nematocide, and as a solvent for resins, gums, and waxes.

Occurrence   No data on EDB concentrations in drinking water were found.

Tissue Distribution   After intraperitoneal injection in rats and mice, EDB was concentrated in the liver, kidneys, and small intestine.

Formation of DNA Adducts   EDB has been shown to bind to DNA in vivo and in vitro. The major adduct formed has been identified as S-[2-(N[7]-guanyl)ethyl]glutathione (Inskeep et al., 1986).

Mutagenicity   EDB is mutagenic at high doses in fungal and bacterial systems. In *Drosophila melanogaster*, gaseous EDB at a concentration as low as 2.3 ppm × hour (0.2 ppm for 11 hours) produced sex-linked lethal mutations (Kale and Baum, 1979).

Carcinogenicity   EDB produces squamous cell carcinoma and adenocarcinoma of the forestomach, liver cancer, and hemangiosarcoma in rats and nasal cavity carcinoma, alveolar carcinoma, bronchiolar carcinoma, hemangiosarcoma, fibrosarcoma, malignant mammary neoplasm, and skin papilloma in mice (NRC, 1986).

Other Health Effects   Humans exposed to EDB have an acute dermal reaction that consists of painful local inflammation, swelling, and blistering. Systemic exposure causes vomiting, diarrhea, and abdominal pain, and in some cases delayed lung damage, depression of the central nervous system, metabolic acidosis, hepatic damage, acute renal failure, and death (NRC, 1986).

Studies in animals have shown acute lethality of EDB in several species. An intraperitoneal dose of 188 mg/kg of body weight in mice produced significant increases in liver and kidney weights and resulted in necrosis in those organs (NRC, 1986).

Summary   EDB is mutagenic, carcinogenic, and toxic in the reproductive system of laboratory animals. It appears to be a direct-acting mutagen in some bacteria and to bind to DNA.

A CONTAMINANT THAT PROBABLY FORMS DNA ADDUCTS

Trichlorfon

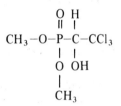

Trichlorfon has been used as a chemotherapeutic agent for schistosomiasis and as an insecticide.

Occurrence   No data on trichlorfon concentrations in drinking water were found.

Tissue Distribution   In experiments with $^{14}CH_3$-labelled trichlorfon, the compound or its metabolites were found in liver, lung, kidney, heart, spleen, and blood. It is also suggested that trichlorfon or a neurologically active metabolite can enter the central nervous system (NRC, 1986).

Formation of DNA Adducts   Trichlorfon can apparently react directly with DNA in vitro, albeit weakly (NRC, 1986).

Mutagenicity   Results of mutagenesis tests and tests for cytogenetic damage in cultured mammalian cells were positive for trichlorfon. Chromatid aberrations, such as sister-chromatid exchanges, have been reported in factory workers exposed to trichlorfon. The genetic risk to the offspring of patients treated with a therapeutic 15-mg/kg dose was estimated to be of the same order of magnitude as the risk associated with 100 mrads of gamma radiation (NRC, 1986).

Carcinogenicity Carcinogenicity studies have produced equivocal results. Mammary tumors of three types were seen in laboratory animals (NRC, 1986).

Other Health Effects Trichlorfon inhibits serum and red blood cell cholinesterases in humans. Humans ingesting high doses developed neurologic dysfunction that appeared to be organophosphorus neuropathy. Decreases in sperm counts and sperm motility have also been noted. Short-lived chromosomal breaks and exchanges and an increase in stable chromosomal alterations were found in some people exposed to trichlorfon. It has also been reported to affect sperm structure (NRC, 1986).

Studies on trichlorfon exposure in animals indicated effects on cholinesterase inhibition similar to those found in humans (NRC, 1986).

Summary The human genetic toxicity risk associated with trichlorfon exposure is not clear. Studies show it has carcinogenic, teratogenic, and reproductive effects in humans or animals. The suggested no-adverse-response level (SNARL) calculated by the NRC Safe Drinking Water Committee is 88 µg/liter for a 70-kg human (NRC, 1986).

CONTAMINANTS THAT POSSIBLY FORM DNA ADDUCTS

Diallate

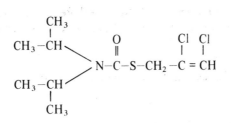

Diallate is an herbicide used in the control of wild oats and weeds in fields of sugar beets, flax, barley, corn, and various other crops.

Occurrence Estimated exposure to diallate for workers during application is approximately 0.5 µg/kg of body weight by inhalation and 980 µg/kg by dermal deposition (NRC, 1986). No data on occurrence in drinking water were found.

Tissue Distribution   No data are available.

Formation of DNA Adducts   Evidence of DNA-adduct formation by dial-late is inconclusive. Diallate causes base-pair mutations and thus might form DNA adducts.

Mutagenicity   Base-pair mutations were produced in *Salmonella typhi-murium* at doses as low as 1 μg/plate. Mutagenic activity depended on metabolic activation by hepatic microsomes (NRC, 1986).

Carcinogenicity   Diallate produced systemic reticulum cell sarcomas in male mice and hepatomas in both sexes of mice (NRC, 1986).

Other Health Effects   In cats, rats, and mice, diallate produced central nervous system excitement that rapidly progressed to clonic convulsions. It produced delayed peripheral neuropathy in hens (NRC, 1986).

Summary   Diallate is mutagenic, carcinogenic, and produces neurotoxic effects. More complete carcinogenicity data is necessary for estimating risk.

Sulfallate

$$C_2H_5 \diagdown \quad \underset{\overset{\|}{\text{S}}}{} \quad \underset{\overset{|}{\text{Cl}}}{}$$

$$N-C-S-CH_2C{=}CH_2$$

$$C_2H_5 \diagup$$

Sulfallate is a carbamate herbicide used for the control of grasses and weeds in fruit and vegetable crops. It is soluble in water up to 100 ppm at 25°C.

Occurrence   No data on the concentration of sulfallate in drinking water were found.

Tissue Distribution   No data are available.

Formation of DNA Adducts   Evidence of DNA-adduct formation by sul-fallate is inconclusive.

Mutagenicity   Sulfallate produces base-pair substitutions in *Salmonella typhimurium* at doses as low as 10 μg/plate in the presence of metabolic activation enzymes (NRC, 1986).

Carcinogenicity Mammary adenocarcinomas in female rats and stomach neoplasms in male rats were observed after the administration of sulfallate; the effect appeared to be dose-dependent. Alveolar and bronchiolar carcinomas and adenomas, with mammary adenocarcinomas, were observed in mice. On the basis of these studies, and assuming consumption of 1 liter of water per day containing sulfallate at 1 μg/liter, the lifetime cancer risk estimated for humans is 1.0×10^{-6}, and the upper 95% confidence estimate of lifetime cancer risk is 1.6×10^{-6} (NRC, 1986).

Other Health Effects Rats fed sulfallate at 250 ppm for 6 months developed eye irritation, tubular nephropathy, and hyperkeratosis of the forestomach (NRC, 1986).

Summary Sulfallate appears to be more carcinogenic than diallate, inasmuch as it produces a broad range of tumors at multiple sites in several organs, whereas diallate produces a limited variety of tumors at a single site (NRC, 1986).

Chloropropanes and Chloropropenes

1,2-Dichloropropane (1,2-DCP)

$$\begin{array}{c} \quad\;\; H\;\; H \\ \quad\;\; |\;\;\; | \\ CH_3-C-C-H \\ \quad\;\; |\;\;\; | \\ \quad\;\; Cl\;\; Cl \end{array}$$

1,2,3-Trichloropropane (1,2,3-TCP)

$$\begin{array}{c} H\;\; H\;\; H \\ |\;\;\; |\;\;\; | \\ H-C-C-C-H \\ |\;\;\; |\;\;\; | \\ Cl\; Cl\; Cl \end{array}$$

cis- **and** *trans-***1,3-Dichloropropene (1,3-DCP)**

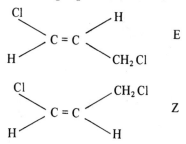

Occurrence   These compounds are generally found in mixtures used as soil fumigants and fungicides, but they have also been used as commercial solvents, drycleaning agents, and in the production of plastics.

No data were found on concentrations of 1,3-DCP, 1,2,3-TCP, and 1,2-DCP in drinking water.

Tissue Distribution   Target organs appeared to be the liver, kidneys, and adrenals in all animals studied (NRC, 1986).

Formation of DNA Adducts   Evidence of DNA-adduct formation is inconclusive.

Mutagenicity   1,2-DCP had equivocal results when tested for mutagenicity in the Ames *Salmonella* test, but was shown to induce sister-chromatid exchanges at 1 mg/ml and chromosomal aberrations at a similar concentration in cultured lymphocytes. 1,3-DCP (with 1% epichlorhydrin added as a stabilizer) produced sex-linked mutations in *Drosophila*. The observed mutagenicity of commercial 1,3-DCP in the Ames test was attributed to the presence of oxygenated and chlorinated degradation products. 1,2,3-TCP proved to be a mutagen in the Ames test at 0.1 μmol/plate, but only in the presence of S9 microsomal extract (NRC, 1986).

Carcinogenicity   Laboratory animals developed multiple hepatomas after exposure to 1,2-DCP. Exposure to 1,3-DCP, given as Telone II (which also contains 1% epichlorohydrin, 2.5% 1,2-DCP, and 1.5% trichloropropene), resulted in tumors of the forestomach and liver nodules in male Fischer-344/N rats and tumors of the urinary bladder, forestomach, and lung in female B6C3F1 mice. 1,3-DCP was shown to be carcinogenic in Swiss mice, but there was insufficient data for a complete assessment. Assuming daily consumption of 1 liter of water containing Telone II at 1 μg/liter, the estimated human lifetime cancer risk is 0.5×10^{-6}, and the upper 95% confidence estimate of lifetime cancer risk is 1.1×10^{-6}. Although no information on carcinogenic studies of 1,2,3-TCP was found, it must be considered a possible carcinogen because of its structural similarity to other compounds displaying carcinogenicity (NRC, 1986).

Other Health Effects   Chloropropanes and chloropropenes cause contact dermatitis, headaches, vertigo, tearing, irritation of mucous membranes, and when ingested, liver damage in humans. The oral LD_{50} is 860 mg/kg of body weight in mice and 2,200 mg/kg in rats (NRC, 1986).

Summary   Most of the studies reported have focused on the toxicity of these compounds in mixtures. 1,2-DCP causes injury in the liver, kidneys, and adrenals, and 1,2,3-TCP has the same target organ toxicity. No other information was found on 1,2,3-TCP, but it is considered carcinogenic because of its structural similarities to the other compounds. 1,3-DCP (as Telone II) is carcinogenic in rats and mice (NRC, 1986).

Di(2-ethylhexyl) Phthalate (DEHP) and Mono(2-ethylhexyl) Phthalate (MEHP)

$$COOCH_2\,CH(C_2H_5)(CH_2)_3CH_3$$
$$COOCH_2\,CH(C_2H_5)(CH_2)_3CH_3$$

$$COOCH_2\,CH(C_2H_5)(CH_2)_3CH_3$$
$$COOH$$

Occurrence   DEHP is used as a plasticizer in plastic products. Polyvinyl chloride pipes can contain up to 30% DEHP. Its concentration in water has been found to range from 5 to 130 ng/liter. DEHP is found in animal products; the highest concentrations have been detected in cheese and milk (35 and 31.4 mg/kg of fat, respectively) (NRC, 1986).

DEHP is easily hydrolyzed in both the environment and the body to MEHP and 2-ethylhexanol. Plasma concentrations of MEHP are always higher than those of DEHP (NRC, 1986).

Tissue Distribution   DEHP is found primarily in liver and adipose tissue.

Formation of DNA Adducts   No evidence of DNA adduct-formation was found.

Mutagenicity   In the Ames *Salmonella* test, DEHP and MEHP were not mutagenic. However, DEHP induced sister-chromatid exchanges in Chinese hamster ovary cells after 24-hour exposures and at 1.0–100 mg/ml was positive in the Syrian hamster embryo cell transformation assay (NRC, 1986). Generally, DEHP and MEHP are considered nonmutagenic.

Carcinogenicity   In light of the observed nonmutagenicity of DEHP, it has been suggested that promotional activity of DEHP might be the mechanism of action whereby carcinogenicity is induced. DEHP-induced S-phase response in mouse hepatocytes to a greater degree than in rat hepatocytes. That suggests that increased cell turnover can account for the carcinogenic activity of DEHP. Assuming daily consumption of 1 liter of water containing the compound at a concentration of 1 mg/liter, the estimated human lifetime cancer risk is 1.2×10^{-7} (NRC, 1986).

Other Health Effects   One study determined that dialysis patients were receiving about 150 mg DEHP per week via leaching from plastics used in their treatment. After 1 month, no structural liver changes were observed; after 1 year, peroxisomes were increased in number. No other data on the effects of DEHP in humans were found. DEHP in the diet of rats at 6,000 and 12,000 ppm and of mice at 3,000 and 6,000 ppm was observed to yield hepatocellular tumors (NRC, 1986).

In male rats, DEHP caused gonadal toxicity that led to reproductive, developmental, and fertility effects (NRC, 1986).

Summary   DEHP is known to alter liver function; other areas of concern for possible toxic effects are reproductive and fertility effects, developmental effects, and cancer.

CONTAMINANTS ON WHICH THERE IS NO EVIDENCE OF DNA-ADDUCT FORMATION

Arsenic

As

Three major sources of arsenic are the smelting of metals, the burning of coal, and arsenic pesticides (for cotton dusting or wood preservation). It is also found in some pigments. Arsenic travels long distances in the atmosphere and precipitates in water. Another source of arsenic in drinking water is its leaching from rocks (Sontag, 1981).

Occurrence   Arsenic concentrations in drinking water can vary from less than 1 ppb to over 600 ppb, depending on the water source. Only 0.4% of samples taken from U.S. public drinking water supplies in a 1970 survey exceeded a total arsenic concentration of 0.01 mg/liter; however, mineral waters can contain 50 times as much arsenic as normal drinking water, and hot springs 300 times as much (NRC, 1977, 1980). The U.S. Environmental Protection Agency drinking water standard is 50 ppb.

Tissue Distribution   Arsenic occurs mainly in the liver, kidneys, spleen, intestinal wall, and lungs (NRC, 1977).

Formation of DNA Adducts   No evidence of DNA-adduct formation has been found. However, arsenic is cocarcinogenic and interferes with DNA repair mechanisms (NRC, 1977).

Mutagenicity   Trivalent arsenic causes aneuploidy. Sister-chromatid exchanges and chromosomal aberrations are noted in cells affected by or exposed to arsenic compounds.

Carcinogenicity   Keratosis, skin cancer, and lung cancer have been observed in persons exposed to arsenic (NRC, 1977).

Other Health Effects   Arsenic primarily affects tissues of the alimentary tract, kidneys, liver, lungs, and epidermis. Its damaging effect on capillaries results in hemorrhage into the gastrointestinal tract, sloughing of mucosal epithelium, renal tubular degeneration, hepatic fatty changes, and necrosis (Merian et al., 1985).

Summary   Arsenic is neurotoxic in humans and animals. Environmental exposure to arsenic in drinking water has been linked in epidemiological studies to an increased incidence of several diseases and cancers (NRC, 1983).

Nitrofen

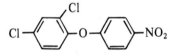

Nitrofen is a contact herbicide used on a variety of food crops to control animal grasses and weeds before and after crops begin to grow. It is activated by sunlight and kills weeds by inhibiting photosynthesis.

Occurrence   No data were found on nitrofen concentrations in drinking water.

Tissue Distribution   Nitrofen concentrates in fatty tissue, and occurs in smaller amounts in other tissues (Hurt et al., 1983).

Formation of DNA Adducts   No data are available.

Mutagenicity   No data are available.

Carcinogenicity   In rats fed nitrofen, metastatic and invasive ductal carcinoma of the pancreas developed. In mice, an increased incidence of hepatocellular carcinomas was observed. On the basis of the no-observed-effect-level (NOEL), and assuming a daily consumption of 1 liter of water containing nitrofen at 1 mg/liter, the estimated human lifetime cancer risk and upper 95% confidence estimate of lifetime cancer risk were 4.4×10^{-5} and 5.6×10^{-5}, respectively (NRC, 1986).

Other Health Effects   A dose-related increase in the weight of the liver, testes, and kidneys was noted in rats fed nitrofen as part of their diet. Elevated liver weights as well as induction of cytochrome P-450 activity levels have been consistent, early signs of low-dose exposure in rodents. Nitrofen produced soft-tissue abnormalities in fetuses of rats, mice, and hamsters (NRC, 1986).

Summary   Nitrofen is teratogenic and carcinogenic in laboratory animals.

Pentachlorophenol

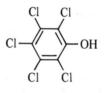

Pentachlorophenol and its salts have been widely used as pesticides. Commercial pentachlorophenol is composed of 88.4% pentachlorophenol, 4.4% tetrachlorophenol, less than 0.1% trichlorophenol, and 6.2% higher chlorinated phenoxyphenols (NRC, 1986).

Occurrence   Pentachlorophenol has been measured in effluent streams at 0.1–10 µg/liter, in river water at 12.3 µg/liter, in surface ponds and drainage water at 1–800 µg/liter, in municipal drinking water at 99 parts per trillion, and in wells at 24 ppm. It has been detected in a variety of wildlife and in fish at 0.35–26 mg/kg of body weight. There has been widespread human exposure to this compound. Pentachlorophenol has been detected in seminal fluid (50–70 mg/kg) of exposed men, in urine (6.3 mg/liter), and in adipose tissue. The plasma of dialysis patients has been shown to contain pentachlorophenol at 15.7–15.8 µg/liter, compared with 15.0 µg/liter in controls (NRC, 1986).

Tissue Distribution   The liver, kidneys, brain, spleen, and fat are major depositories in humans (NRC, 1986).

Formation of DNA Adducts   No evidence of DNA-adduct formation has been found.

Mutagenicity   In the mouse spot test, weak mutagenic activity was observed. Pentachlorophenol was mutagenic in yeast assay, and chromosomal abnormalities were observed in *Vicia faba* seedlings, but no increases in chromosomal aberrations were observed in workers with increased concentrations of pentachlorophenol in serum and urine (NRC, 1986).

Carcinogenicity   Carcinogenicity studies have not provided a basis for complete evaluation.

Other Health Effects   Acute exposure of laboratory animals to pentachlorophenol results in vomiting, hyperpyrexia, and increases in blood pressure, respiration rate, and heart rate. Chronic studies have shown alterations in liver structure at all doses when rats ingested 20–500 ppm of pentachlorophenol over an 8-month period. Immunotoxicity and neurotoxicity were observed in laboratory animals given pentachlorophenol, and developmental effects occurred in the litters of female rats and hamsters dosed during gestation (NRC, 1986).

Summary   The metabolism of pentachlorophenol is generally similar among mammalian species. However, inadequate characterization of the pentachlorophenol used in studies has led to uncertainty as to whether observed effects were due to pentachlorophenol or to contaminants. Pentachlorophenol can be absorbed through the skin, by inhalation, and by ingestion. In animals, it has resulted in fetotoxicity and damage to the liver, kidney, and central nervous system; hematologic and immune systems have been affected. In humans, it is neurotoxic. It affects the immune system, liver, and kidneys, and produces hematologic disorders. Some studies have reported production of aplastic anemia and malignancy.

REFERENCES

Hurt, S. S., J. M. Smith, and A. W. Hayes. 1983. Nitrofen: A review and perspective. Toxicology 29:1–37.

Inskeep, P. B., N. Koga, J. L. Cmarik, and F. P. Guengerich. 1986. Covalent binding of 1,2-dihaloalkanes to DNA and stability of the major DNA adduct S-[2-(N^7-guanyl)ethyl]glutathione. Cancer Res. 46:2839–2844.

Jeffrey, A. M., K. Grzeskowiak, I. B. Weinstein, K. Nakanishi, P. Roller, and R. G. Harvey.

1979. Benzo(*a*)pyrene-7,8-dihydrodiol 9,10-oxide adenosine and deoxyadenosine adducts: Structure and stereochemistry. Science 206:1309–1311.

Kale, P. G., and J. W. Baum. 1979. Sensitivity of *Drosophila melanogaster* to low concentrations of gaseous mutagens. II. Chronic exposures. Mutat. Res. 68:59–68.

Kapp, R. W., D. J. Picciano, and C. B. Jacobsen. 1979. Y-Chromosomal nondisjunction in dibromochloropropane-exposed workmen. Mutat. Res. 64:47–51.

Kato, Y., K. Sato, O. Matano, and S. Goto. 1980. Alkylation of cellular macromolecules by reactive metabolic intermediate of DBCP. Nippon Noyaku Gakkaishi (J. Pest. Sci.) 5:45–53.

McCann, J., E. Choi, E. Yamasaki, and B. N. Ames. 1975. Detection of carcinogens as mutagens in the Salmonella/microsome test: Assay of 300 chemicals. Proc. Natl. Acad. Sci. USA 72:5135–5139.

Merian, E., R. W. Frei, W. Hardi, and C. Schlatter, eds. 1985. Carcinogenic and Mutagenic Metal Compounds. Current Topics in Environmental and Toxicological Chemistry, Vol. 8. New York: Gordon and Breach.

Miller, M. J., D. E. Carter, and I. G. Sipes. 1982. Pharmacokinetics of acrylamide in Fisher-344 rats. Toxicol. Appl. Pharmacol. 63:36–44.

Moore, M. M., A. Amtower, C. Doerr, K. H. Brock, and K. L. Dearfield. 1987. Mutagenicity and clastogenicity of acrylamide in L5178Y mouse lymphoma cells. Environ. Mutagen. 9:261–267.

NRC (National Research Council). 1977. Drinking Water and Health, Vol. 1. Washington, D.C.: National Academy of Sciences. 939 pp.

NRC (National Research Council). 1980. Drinking Water and Health, Vol. 3. Washington, D.C.: National Academy Press. 415 pp.

NRC (National Research Council). 1982. Drinking Water and Health, Vol. 4. Washington, D.C.: National Academy Press. 299 pp.

NRC (National Research Council). 1983. Drinking Water and Health, Vol. 5. Washington, D.C.: National Academy Press. 157 pp.

NRC (National Research Council). 1986. Drinking Water and Health, Vol. 6. Washington, D.C.: National Academy Press. 457 pp.

Rigdon, R. H., and J. Neal. 1966. Gastric carcinomas and pulmonary adenomas in mice fed benzo(*a*)pyrene. Tex. Rep. Biol. Med. 24:195–207.

Rigdon, R. H., and J. Neal. 1969. Relationship of leukemia to lung and stomach tumors in mice fed benzo(*a*)pyrene. Proc. Soc. Exp. Bio. Med. 130:146–148.

Sasaki, Y. F., H. Imanishi, M. Watanabe, A. Sekiguchi, M. Moriya, Y. Shirasu, and K. Tutikawa. 1986. Mutagenicity of 1,2-dibromo-3-chloropropane (DBCP) in the mouse spot test. Mutat. Res. 174:145–147.

Shelby, M. D., K. T. Cain, C. V. Cornett, and W. M. Generoso. 1987. Acrylamide: Induction of heritable translocation in male mice. Environ. Mutag. 9:363–368.

Shubik, P., G. Pietra, and G. Della Porta. 1960. Studies of skin carcinogenesis in the Syrian golden hamster. Cancer Res. 20:100–105.

Solomon, J. J., J. Fedyk, F. Mukai, and A. Segal. 1985. Direct alkylation of 2′-deoxynucleosides and DNA following in vitro reaction with acrylamide. Cancer Res. 45:3465–3470.

Sontag, J. M., ed. 1981. Carcinogens in Industry and the Environment. Pollution Engineering and Technology, Vol. 16. New York: Marcel Dekker. 761 pp.

Teramoto, S., R. Saito, H. Aoyama, and Y. Shirasu. 1980. Dominant lethal mutation induced in male rats by 1,2-dibromo-3-chloropropane (DBCP). Mutat. Res. 77:71–78.

U.S. EPA (Environmental Protection Agency). 1984. Health Assessment Document for Chromium. Final Report. EPA-600/8-83-014F. NTIS PB85-115905. Research Triangle Park, N.C.: U.S. Environmental Protection Agency.

Glossary

ACTIVATING ENZYMES Enzymes or enzyme complexes associated with activation of a xenobiotic compound. Such enzymes typically oxidize the compound to reactive intermediates or detoxify the compound.

ADDUCT A chemical addition product. For example, when the mutagenic alkylating agent ethyl methanesulfonate reacts with DNA, any of the normal bases in DNA (i.e., adenine, thymine, guanine, and cytosine) may be converted into adducts, such as N7-ethylguanine and N3-ethylcytosine. Adducts may also be formed between bases on the phosphate backbone of DNA. An example is the formation of pyrimidine dimers upon exposure to the physical agent ultraviolet light. Most chemical agents that react with DNA can also form adducts with proteins; however, the extent of the reaction with each often differs greatly.

ALKYLATING AGENT A substance that causes the addition of an alkyl group to an organic compound; according to the number of reactive groups they contain, alkylating agents are classified as monofunctional, bifunctional, or polyfunctional; many alkylating agents are mutagenic to humans.

ALKYLATION The addition of alkyl groups, such as methyl or ethyl groups, to another chemical; for example, the mutagen ethyl methanesulfonate (EMS) adds ethyl groups to DNA, forming adducts such as N7-ethylguanine; EMS is said to be an alkylating agent and to alkylate DNA.

79

ANTIGEN   A substance capable, under appropriate conditions, of inducing a specific immune response and of reacting with the products of that response.

ANTISERUM   An antibody-containing serum generated by an animal or human as an immunologic response to infection.

8-AZAGUANINE   An analogue of the normal DNA and RNA purine base guanine; selection for resistance to the toxic effects of 8-azaguanine is the basis for several mutation-detection systems.

BASELINE ADDUCTS   Endogenous adducts found on DNA from "unexposed" organisms or individuals. The source of these adducts is unknown.

BINDING   Formation of adducts by chemical linkage between an exogenous compound and a normal cellular macromolecule.

BIOLOGIC DOSIMETER   The use of a cellular molecule that undergoes a change following a reaction with a specific agent or class of agents as an estimate of dose to the target tissue or as a marker of exposure or effect.

BIOLOGIC END POINT   Target tissue; the site of toxic action.

BIOTRANSFORMATION   The series of chemical alterations of a compound occurring within the body, as by enzymatic activity.

CARCINOGEN   A chemical or physical agent which causes an increase in tumor rate in exposed organisms or individuals.

CENTROMERE   A specialized part of a chromosome that attaches to a spindle fiber during cell division.

CHROMATID   One of the identical longitudinal halves of a chromosome, sharing a common centromere with a sister chromatid; produced by the replication of a chromosome during interphase. (*See* **SISTER-CHROMATID EXCHANGE**)

CHROMOSOME   A nucleoprotein structure, generally more or less rodlike during nuclear division; a physical structure that bears genes; each species has a characteristic number of chromosomes.

CHRONIC ANIMAL BIOASSAY   Long-term (90 days to lifetime) exposure study, usually conducted in small rodents (mice and rats), to determine, in the context of this report, the carcinogenicity of a chemical or physical agent.

CODING SEQUENCE The region of a gene (DNA) that encodes the amino acid sequence of a protein.

COVALENT BINDING A stable chemical bond formed between two atoms in which electrons are shared by both.

CROSS-LINKING The joining of two DNA strands or a DNA strand with a protein strand by a bifunctional or polyfunctional alkylating agent. An **interstrand** cross-link binds two separate strands together; an **intrastrand** cross-link joins different parts of the same DNA strand.

CYTOSINE DEAMINATION The removal of the NH_2 group from the C4 atom of cytosine, converting the cytosine to uracil.

DNA Deoxyribonucleic acid, the genetic material of every cell.

DNA ALTERATIONS Changes in the nucleotide sequence or the structural integrity of the normal DNA molecule of a cell or organism.

DNA DAMAGE Any modification of DNA that alters its normal structure, its coding properties, or its normal function in replication or transcription.

DNA LESIONS DNA alterations that form the basis of mutation.

DEPURINATION The process of removal of a purine base, either guanine or adenine, from DNA.

DEPYRIMIDINATION The process of removal of a pyrimidine base, either thymine or cytosine, from DNA.

DETOXIFICATION The metabolic conversion of a substance into another substance of lower toxicity; the mammalian liver is an important site of detoxification processes. (*See also* **METABOLIC ACTIVATION** *and* **TOXIFICATION**)

DEVELOPMENTAL EFFECTS Embryonic (fetal) growth retardation, malformations, or death produced by exposures during embryonic stages of development, usually at levels that do not induce severe toxicity in the mother.

DIPLOID An organism or cell having two complete sets of chromosomes, with each set typically of a different parental origin.

DOMINANT LETHAL MUTATION A mutation of the germ cell that results in embryonic or fetal death.

DOSE   Used loosely as equivalent to quantity or extent of exposure. Specifically, as **DOSE TO THE TARGET TISSUE**, the amount of material, expressed as time integral of concentration, that reaches a biologically significant target; **BIOLOGICALLY EFFECTIVE DOSE**, the number of adducts per nucleotide or per unit of protein.

DOSIMETRY   Scientific determination of amount, rate, and distribution of a dose (chemical or physical) to a particular target.

ELECTROPHILE   An agent with an affinity for a pair of electrons that bonds with a substance (i.e., with a nucleophile) offering a pair of electrons; because there are many nucleophilic sites in DNA, electrophiles can react with DNA to produce a variety of adducts; many metabolically activated mutagens and carcinogens or their metabolites are electrophilic.

ENZYMATIC REPAIR SYSTEMS   Cellular enzymes capable of recognizing and repairing altered DNA templates.

EXCISION REPAIR   The enzymatic removal from DNA of an altered base or nucleotide, followed by resynthesis and rejoining of the DNA. Various different enzymatic pathways for excision repair are recognized.

EXOCYCLIC   Refers to chemical groups not in the basic ring structures of the bases in DNA. For example, the NH_2 group on the C4 atom of cytosine is such a group.

EXPOSURE   In the context of this report, amount of material ingested, inhaled, or otherwise received by an organism. (*See also* **DOSE**)

FEMTOMOLE 10^{-15} moles.

GENOME   A complete set of chromosomes or of chromosomal genes.

GENETIC TOXICITY   The capacity to cause an adverse effect on a genetic system, including mutagenesis and other indicators of genetic damage.

GERM CELL   Any cell that will eventually differentiate into a mature sex cell (sperm or oocyte).

HAPTEN   The portion of an antigenic molecule or complex that determines its immunologic specificity.

HAZARD IDENTIFICATION   Determination of the potential for an agent to produce toxicity.

HEPATOCARCINOGEN   A chemical that causes cancer in the liver.

HEPATOCELLULAR CARCINOMA   A cancer pertaining to or affecting liver cells.

HEPATOCYTE   A parenchymal liver cell.

HERITABILITY   The ability to be passed on from one cell generation to another or from one individual to an offspring.

HERITABLE TRANSLOCATION   A stable rearrangement of the position of chromosomal segments that leads to successful chromosomal replication and progeny that carry the translocation.

HPRT   Hypoxanthine-guanine phosphoribosyl transferase (also called HGPRT); an enzyme involved in the utilization of the purine bases hypoxanthine and guanine in mammalian cells (there are related enzymes in submammalian species; mutants that lack HPRT are resistant to the toxic effects of the guanine analogues 8-azaguanine and 6-thioguanine, which can therefore be used to select HPRT mutants and form the basis of several mutation-detection systems.

"HOT SPOT"   A DNA site or sequence that is particularly susceptible to the formation of alterations that result in mutations.

IMMUNOGEN   Any substance capable of eliciting an immune response.

INTRAPERITONEAL   Referring to the abdominal region within the peritoneum.

LOCUS   The position that a particular gene occupies in a chromosome.

METABOLIC ACTIVATION   The metabolic conversion of a promutagen into a mutagen—as an aspect of toxification; the possibility that a chemical may undergo toxification in vivo provides the rationale for using S-9 mixtures or other metabolic activation systems with many in vitro genetic toxicity tests. (*See also* **ACTIVATING ENZYME** *and* **DETOXIFICATION**)

METHYLATING AGENT   A chemical that can transfer a methyl (CH_3) group to another compound.

MISCODING   An alteration in DNA that changes the readout of information from the usual amino acid to a different amino acid.

MOLECULAR DOSIMETRY   The measurement of chemical dose at a particular target organ or tissue by examination of configurational or chemical changes in cellular molecules.

MUTAGEN   An agent that causes mutation.

MUTATION   A permanent change in genotype other than one brought about by genetic recombination.

NONINVASIVE   In reference to analytical techniques requiring a test specimen, not involving injury to the subject.

NUCLEOPHILIC SITE   Within a chemical structure, a region with a negative charge that attracts positively charged chemical groups, or donates electrons. (*See also* **ELECTROPHILE**)

NUCLEOTIDE   The monomeric unit of polynucleotide polymers known as nucleic acids; consists of three components—a ribose or a 2-deoxyribose sugar, a pyrimidine or purine base, and a phosphate group—each of which exists as a phosphate ester of the *N*-glycoside of the nitrogenous base.

NUCLEOTIDE EXCISION   Repair by the enzymatic removal of a nucleotide or oligonucleotide containing an altered base followed by synthesis and rejoining of the DNA. (*See* **EXCISION REPAIR**)

PHARMACODYNAMIC   Pertaining to the biochemical processing and physiologic effects of xenobiotic compounds within a living organism.

PHARMACOKINETICS   The movement in time of xenobiotic compounds in the body, including the processes of absorption, distribution, localization in tissues, biotransformation, and excretion.

PROMOTER   A chemical compound that advances the process of carcinogenesis begun by other agents.

PROMUTAGEN   An agent that can become a mutagen if metabolically activated.

PROTAMINE   A small, basic protein that is rich in the amino acid argenine and contains a positive charge. In late stages of spermatid maturation, protamines neutralize the negative charges on DNA by replacing the usual uncharged chromosomal proteins (histones) and thus facilitate DNA folding and packaging.

PURINES   Two of the bases found in DNA belong to this chemical class. They are guanine (G) and adenine (A). They are linked to the C1 position of deoxyribose, the sugar found in the DNA backbone.

PYRIMIDINE   The other two bases in DNA belong to this chemical class. They are cytosine (C) and thymidine (T). They are linked to the C1 position of deoxyribose, the sugar found in the DNA backbone. The usual base pairs in DNA are G–C and A–T.

PYRIMIDINE DIMERS   Adjacent pyrimidine bases in DNA that are joined by a cyclobutane ring as a result of short wavelength UV irradiation.

RADIOLABELING   The process of incorporating a radioactive isotope of an atom into a larger chemical structure to produce a marker.

RECESSIVE LETHAL MUTATION   A gene alteration which, when homozygous, results in a mutant phenotype, but which, when heterozygous, results in a normal or near normal phenotype.

REPLICATION   In the synthesis of new DNA from preexisting DNA, the formation of replicas from a model or template; the process by which genes (hereditary material; DNA) duplicate themselves.

RISK ASSESSMENT   Includes four components: hazard identification, exposure assessment, dose-response assessment, and characterization of risk at projected levels and patterns of exposure. **Qualitative** risk assessment involves only the first two, and sometimes a limited amount of the third, whereas **quantitative** estimates of human risk include all four components and involve more uncertainties and less consensus about the most appropriate animal models, data sets, and conversion factors to use in calculation. Nevertheless, there are compelling arguments favoring the use of animal data for quantitative risk assessments.

S-9   A metabolic activation mixture (i.e., microsomal and cytosolic enzymes) derived from a mammalian liver homogenate. S-9 is used with many in vitro genetic-toxicity tests to provide for the conversion of promutagens into mutagens. (*See also* **ACTIVATING ENZYME**)

SENSITIVITY   The proportion of test cells or animals that are positive in the system being evaluated; the capacity of a test to detect small increases in toxic effects.

SINGLE-STRAND BREAK A break occurring in the phosphate backbone of one strand of a double-stranded DNA molecule.

SISTER-CHROMATID EXCHANGE The exchange of segments between the two chromatids of a chromosome. (*See also* **CHROMATID**)

SPECIFICITY The quality of affecting only certain organs or tissues, or reacting only with certain substances.

SPERMATIDS Immature male germ cells that are nearly fully differentiated as sperm cells.

SPERMATAZOA Fully mature sex cells in the male.

SPERMIOGENIC Referring to those germ cell stages in the male from meiosis onward, in which spermatids and spermatozoa are differentiating.

TEMPLATE In genetics, a strand of DNA that specifies the synthesis of a strand of ribonucleic acid (RNA) complementary to itself or of messenger RNA that in turn serves as a template for the synthesis of proteins.

6-THIOGUANINE An analogue of the purine base guanine, which is a normal component of DNA and RNA; selection for resistance to the toxic effects of 6-thioguanine is the basis of several mutation-detection systems.

THYMIDINE KINASE (TK) An enzyme involved in the utilization of the nucleotide thymidine (which ultimately becomes part of the structure of DNA); catalyzes the phosphorylation of thymidine to thymidine monophosphate; mutants that lack TK are resistant to the toxic effects of several thymidine analogues, including bromodeoxyuridine and trifluorothymidine; selection of these drug-resistant mutants provides the basis of several mutation-detection systems, most notably in mammalian cells.

TIME-WEIGHTED AVERAGE An average exposure concentration calculated from intermittent samplings of exposure concentrations.

TOXICANT Any substance that, through its chemical action, causes adverse effects in living organisms.

TOXIFICATION The metabolic conversion of a substance into another substance that has greater toxicity; sometimes occurs as a consequence of processes that are usually associated with detoxification. (*See also* **METABOLIC ACTIVATION**)

TRANSCRIPTION   The synthesis of RNA on a DNA template in such a way that the sequence of nucleotides in the RNA is the complement of the sequence of nucleotides in DNA.

VALIDATION   The process by which the consistency of a particular test is determined; the concordance of results of a test in question and previously established tests for a representative sample of chemicals is evaluated.

XENOBIOTIC   Originating outside or caused by factors outside the organism; pertaining to that which is not a normal physiologic part of an organism.

Biographical Sketches

SUBCOMMITTEE ON DNA ADDUCTS

DAVID J. BRUSICK received his Ph.D. in microbial genetics from Illinois State University in 1970 and did postdoctoral research as a National Academy of Sciences research associate at the Food and Drug Administration's Genetic Toxicology Branch. A past president of the U.S. Environmental Mutagen Society (1978–79), Dr. Brusick is adjunct professor of microbiology and genetics at the schools of medicine of both Howard and George Washington universities. He is the author of numerous scientific publications, including a textbook, *Principles of Genetic Toxicology*. He was a committee member contributing to the NRC report *Toxicity Testing: Strategies to Determine Needs and Priorities* for the National Toxicology Program and has also served on numerous other NRC committees. He is a member of the International Commission for the Protection against Environmental Mutagens and Carcinogens and a member of the Steering Committee for the Environmental Protection Agency's Gene-Tox Program. Dr. Brusick's interests include basic and applied research in mutagenic and carcinogenic mechanisms and the application of biotechnology techniques to the development of safety testing methods.

GAIL T. ARCE is a genetic toxicologist at the Haskell Laboratory for Toxicology and Industrial Medicine, E. I. du Pont de Nemours & Company, in Newark, Delaware. She received her Ph.D. from Yale University in 1978 from the Department of Molecular Biophysics and Biochemistry and did postdoctoral work at New York University's Department of Environmental

89

Medicine and Columbia University's Institute of Cancer Research. Her research has focused on the evaluation of DNA-adduct dosimetry in in vitro mutation and transformation assays.

JOHN C. BAILAR III is a statistician and physician at McGill University in Montreal. Since 1980 he has been a statistical consultant for the *New England Journal of Medicine*. His research interests center on the processes of health risk assessment, whether they are applied to chemicals, radiation, microorganisms, or other hazards. He also has a strong interest in scientific communication, and in 1987–88 served as the president of the Council of Biology Editors. He has been a member of many National Academy of Sciences/National Research Council studies concerned with health risks.

RAMESH C. GUPTA holds degrees from Agra, Meerut, and Roorkee Universities in India. He is an associate professor in the Department of Pharmacology at Baylor College of Medicine in Houston, Texas. Conducting research in sensitive techniques for sequencing of RNA, he is also one of the developers of the ^{32}P-postlabeling assay for DNA adducts. His research interests include DNA damage and repair in animal and human cells. Dr. Gupta is a member of the American Society of Biological Chemists, the American Society of Cancer Research, the American Society for Cell Biology, and the American Association for the Advancement of Science.

ROBIN HERBERT is a Charles A. Dana Foundation Fellow in Environmental Epidemiology and an instructor in the Division of Environmental and Occupational Medicine of the Department of Community Medicine at Mount Sinai Medical Center in New York, New York. An internist and specialist in occupational medicine, Dr. Herbert's principal area of interest is in the use of biologic markers of exposure to carcinogens in occupationally and environmentally exposed populations. She is currently studying DNA adducts and other markers of exposure among roofing workers exposed to polycyclic aromatic hydrocarbons. Dr. Herbert is a member of the American College of Physicians, the Society for General Internal Medicine, and the American Public Health Association.

PAUL H. M. LOHMAN is a professor and director of the laboratory of Radiation Genetics and Chemical Mutagenesis, State University of Leiden, The Netherlands. Dr. Lohman is an expert is the field of DNA repair and the relation between the induction of DNA damage, DNA repair, and mutagenesis in cells of mammalian origin. Currently, he is heading one of the largest research institutions in Europe in the filed of environmental mutagenesis and genetic toxicology. He is scientific secretary of the International

Commission for Protection against Environmental Mutagens and Carcinogens and a past president of the European Environmental Mutagen Society.

CAROL W. MOORE received degrees from Ohio State University and Pennsylvania State University and performed postdoctoral research at the University of Rochester School of Medicine. She received a fellowship to conduct research in the Interdisciplinary Programs in Health and Environmental Health Policy and Management Program from the Harvard University School of Public Health and School of Business. Dr. Moore is currently associate professor of microbiology at City University of New York Medical School, Sophie Davis School of Biomedical Education—City College, New York City. She has conducted research in genetics, molecular biology, genetic toxicology, cancer biology, radiobiology and biochemistry, including the genetic control of cellular responses in yeast and human cells to radiation and the radiomimetic bleomycins.

ROBERT F. MURRAY is a graduate of the University of Rochester School of Medicine, an internist whose subspecialty is medical genetics. A military tour of duty in the U.S. Public Health Service at the National Institutes of Health sparked an interest in genetic markers which might indicate inherited susceptibility to disease. He received a master's degree and a fellowship in medical genetics at the University of Washington in Seattle. After joining the Faculty of Medicine at Howard University in Washington, D.C., he continued studies of developmental variation in human liver alcohol dehydrogenase and genetic markers indicating the clonal origins of breast tumors. He also became involved in programs of genetic screening, counseling and prenatal diagnosis (with special emphasis on sickle cell disease), and the use of genetic markers to identify individuals highly susceptible to potentially toxic compounds in the work environment, as well as the effects of these compounds on the human genome. A major research interest has been the study of the psychological aspects of genetic counseling. He is currently chief of the Division of Medical Genetics in the Department of Pediatrics and Child Health, directing an active program of teaching and patient service, and chairman of the Graduate Department of Genetics and Human Genetics. He is an active member of the Institute of Medicine and he has served on its governing council and also on several NRC and IOM task forces and working groups.

MIRIAM C. POIRIER is a research chemist in the Laboratory of Cellular Carcinogenesis and Tumor Promotion at the National Cancer Institute, NIH. Her graduate studies were carried out at the McArdle Laboratories at the University of Wisconsin, and the Department of Biology at the Catholic University of America. She was among the first to elicit antisera specific for carcinogen-DNA adducts and investigate mechanisms of chemical carcinogenesis utilizing quantitative immunoassays and immunohistochemistry. For development of this methodology she received the NIH Merit Award. More recently Dr. Poirier has pioneered efforts to validate the use of immunoassays for human exposure biomonitoring. She has served in an advisory capacity to the Harvard School of Public Health, the Department of Health and Human Services Panel on Application of Biologic Markers in Risk Assessment, and the Health Effects Institute.

GARY A. SEGA received graduate degrees from the University of Texas and Louisiana State University. He has been a research staff member in the Biology Division of Oak Ridge National Laboratory for the past 15 years. Dr. Sega's principal research interest is studying the molecular mechanisms that give rise to mutations in mammalian germ cells, including chemical binding to DNA and proteins, and DNA repair in the germ cell. He is a member of the Environmental Mutagen Society and is presently a member of the editorial board of *Molecular and Environmental Mutagenesis.*

RICHARD B. SETLOW was educated in the field of physics (Ph.D., Yale University, 1947). He is a senior biophysicist at Brookhaven National Laboratory on Long Island and is the Laboratory's Associate Director for Life Sciences. His research deals with DNA repair mechanisms in numerous biological systems, the roles of such repair mechanisms in ameliorating the effects of carcinogenic chemicals, variations in repair among people, and the association of such mechanisms with aging. He is a member of the National Academy of Sciences.

JAMES A. SWENBERG is head of the Department of Biochemical Toxicology and Pathobiology at the Chemical Industry Institute of Toxicology and an adjunct professor of pathology at the University of North Carolina and Duke University. Dr. Swenberg is on the editorial boards of several cancer- and toxicology-related journals; served as a member and chairperson of the National Toxicology Program Board of Scientific Counselors; is a member of the EPA FIFRA Science Advisory Panel; and the Board of Scientific Counselors, Division of Biometry and Risk Assessment of the National Institute for Environmental Health Sciences. Dr, Swenberg's research interests address the role of DNA adducts, repair and replication in carcinogenesis, experimental neuroncology and the scientific basis of quantitative risk assessment. He is a diplomat in the American College of Veterinary Pathologists.

PART II
Mixtures

Executive Summary

Humans live in a complex environment and are often exposed to sequences or mixtures of toxic materials. The science of dealing with the toxicity of mixtures is relatively new and is still developing. Most experimental data relate to the toxic effects of exposures to single materials, yet exposure to two or more toxic materials might produce greater deleterious effects than would be anticipated from knowledge of the effects of each of the materials considered separately. Even as few as two materials can be present in infinite dose combinations, and relative doses might affect toxicity of the mixture. Experimental strategies for testing a manageable sample of the infinite combinations are obviously needed. Even if the proportions of the constituents are not of consequence (i.e., if only presence or absence is important), many combinations can be made from a few constituents. For example, if each mixture were treated as a separate material, testing all combinations of toxicants would be physically and economically impossible. Testing all possible mixtures of 10 substances (i.e., 2 at a time, 3 at a time, etc.) would require more than 1,000 tests, even if each substance were tested at only one dose (present or absent).

An awareness of these problems and the recognition that the biological mechanisms of toxicity of even single materials are often not known led the Environmental Protection Agency (EPA) to ask the Board on Environmental Studies and Toxicology of the National Research Council, through its Safe Drinking Water Committee, to convene a workshop (1) to review the scientific and experimental issues associated with estimating the toxicity of mixtures in drinking water, (2) to suggest possible modifications of the current approaches, and (3) to recommend subjects for future research. This report is the product of the workshop. Because the study of the toxicity of mixtures is relatively new, the Subcommittee on Mixtures, which prepared the report, confined its considerations to a small number of issues.

APPROACHES FOR PROBLEM RESOLUTION

Several possibilities exist for attacking the problems of testing (and regulating) mixtures. Research on mechanisms could lead to theoretical modeling that would exploit data on the toxicity of single materials; that is a desirable long-term goal. In the absence of knowledge of mechanisms and of accepted models for combining the toxicities of single materials, some mixtures will have to be tested. If the numbers of such mixtures can be kept small, testing might become both operationally and economically possible.

GROUPING CHEMICALS FOR ESTIMATION OF COMBINED RISK

A prudent first step in assessing and managing the health risks associated with exposure to mixtures in drinking water is to reduce the apparent number of mixtures by developing appropriate methods for grouping the substances found in water. This report examines two such groups based on similarities of biologic effects: volatile organic chemicals (VOCs), of which several are suspected human carcinogens, and organophosphorus compounds and carbamates, which inhibit acetylcholinesterase. In addition, this report suggests some ways to make other logical classifications and thus reduce the testing needed.

Except for the trihalomethanes (a class of volatile organic chemicals that are by-products of disinfection), the EPA Office of Drinking Water has promulgated standards only for single components of drinking water. It is unlikely that each agent present in drinking water will act in biologic isolation of every other agent present. Where the mechanisms of toxicity of two or more toxicants are the same or similar, exposure to several materials, each at a below-threshold dose (i.e., a dose with a zero response), could amount to exposure to an above-threshold dose and produce a response.

EPA's general guidelines for evaluating the health risk associated with exposure to a chemical mixture recommend that, if data are not available on the complete mixture or a toxicologically similar mixture, a dose-additive method be used to define a "hazard index" (in effect, an exposure index) based on measured concentrations and reference doses. The subcommittee recommends using a modified hazard index that sums similar toxicities, incorporates multiple toxic manifestations, and suggests, under some circumstances, the use of an uncertainty factor to allow for possible synergisms.

The subcommittee reviewed earlier schemes for the arbitrary grouping of substances in its efforts to devise some classification schemes that could facilitate control of mixtures in drinking water. In light of EPA's general guidelines, the subcommittee suggested the following as possible ways for EPA to group substances for assessing their combined risk when they are found in drinking water. Groupings that are more rational for regulatory purposes might be found.

- Substances can be grouped *according to carcinogenicity*. According to the currently preferred dose-extrapolation models, additivity of response or risk can usually be assumed for low-dose exposure to a mixture of carcinogens (at doses with relative risks of less than 1.01). The subcommittee cautions that additivity might not apply for carcinogens at high doses, when some dose extrapolation models are considered, or when one agent is a "pure" initiator and another a promoter.
- Substances can be grouped *according to other toxic end points* (such as specific organ toxicity, peripheral nerve damage, etc.). It is likely that not all members of a group will affect an end point via the same mode of action.
- When toxic end points are unknown, substances can be grouped *on the basis of their transformation into similar metabolites* with similar reactivity and stability, although it must be kept in mind that their toxic end points might differ.
- Substances can be grouped *according to structural properties*. For toxic materials that fall into any of these groups, a toxic-equivalence approach that estimates the combined toxicity of the members of a single chemical class on the basis of the toxicity of a representative of the class has substantial appeal. In this approach, one estimates the potencies of the contaminants belonging to a given class relative to the potency of a representative of the class. (That is already done for some classes of compounds, such as polycyclic aromatic hydrocarbons, dibenzo-*p*-dioxins, and dibenzofurans). Toxic doses can then be combined according to some weighting procedure using a dose-additive model.

USING PHYSIOLOGICALLY BASED PHARMACOKINETIC MODELING

The new field of physiologically based pharmacokinetics warrants attention as a first step in applying existing knowledge of mechanisms of toxicity. (In the realm of toxicity modeling, the term "toxicokinetics" might be more appropriate.) Unfortunately, little is known about how pharmacokinetic variables are affected by simultaneous or sequential exposure to multiple chemicals. Improved understanding and modeling of the pharmacokinetics of mixtures should lead to improved methods for estimating the risks associated with exposure to multiple chemicals in drinking water. Development of appropriate pharmacokinetic models will require considerable theoretical and experimental work.

STATISTICAL APPROACHES FOR REDUCING EXPERIMENTATION (RESPONSE-SURFACE DESIGNS AND FRACTIONAL FACTORIALS)

Response-surface methods are mathematical-statistical techniques that permit the estimation of the effects of each component of a mixture and the effects of interactions (i.e., departures from additivity) on the basis of a small number of experimental points. The 2^k factorial approach permits the estimation of possible interactions: each of the k factors (elements in the mixture) is assumed to be present at two concentrations (one of which may be zero), and each concentration of each factor is combined with each concentration of every other factor. Usually, some interactions are of less interest, and that permits the use of fractional factorial designs, which require still fewer experimental groups. The availability of modern computer graphics has permitted the plotting of the data, both observed and fitted, in ways that lead to easier identification of nonadditivity, if there is any.

ASSESSING EXPOSURE

To develop a risk assessment of a drinking water mixture, exposure needs to be estimated. Drinking water is often not the sole source of exposure to many of its contaminants. Exposures due to contact with other media and by other routes must be considered, because they have the potential for raising total body burden to levels that could be of concern and for providing substances critical for synergistic interaction with waterborne substances. Inhalation of volatilized drinking water contaminants during cooking, showering, or other activities is another route by which water can contribute additional exposure, as is dermal contact in swimming and bathing.

The subcommittee considers it likely that, if a simple analytic process were developed to provide a summary measure of an entire class of toxicologically similar constituents in drinking water, it would also detect other, potentially confounding constituents in the water.

APPLICATIONS—PROBLEMS OF EXPOSURE

On a practical level, ambient exposures to mixtures usually involve low concentrations of the constituents. At concentrations that yield small increases in relative risks, additive and multiplicative responses are essentially indistinguishable, and additivity is a satisfactory first approximation. For example, several organophosphorus and carbamate chemicals inhibit acetylcholinesterase, and their joint action is assumed to be the sum of their separate actions on this end point.

The additive approach might need to be modified by incorporating an uncertainty factor (for possible synergism), which would depend in part on the information available and the concentrations of the contaminants. If a great deal of toxicologic information is available on the individual contaminants, if toxic interactions are not likely (on the basis of the knowledge available), or if the concentrations of the contaminants are low, the uncertainty factor might be set at 1 (producing an assumption of simple additivity). If less is known about the toxicity of individual components and the concentrations of the contaminants are higher, the uncertainty factor might be set at 10. The greater the uncertainty (because of lack of information) and the higher the concentrations of the contaminants, the larger the uncertainty factor required to provide an adequate margin of safety. For example, synergism and antagonism among some anticholinesterases are known to depend on interference with or competition for metabolic mechanisms of detoxification or activation of the anticholinesterases or their precursors. One might predict that synergism could occur only when the dosages are high enough for metabolic detoxification to be rate-limiting with respect to toxicity. The dosages necessary could be lower when other substances inhibit critical pathways of detoxification, particularly if the inhibition is of a noncompetitive type. At the (low) concentrations of anticholinesterase compounds found in drinking water, it is probably safe to assume no more than additivity of effect. For such concentrations, the uncertainty factor would be 1.

MIXTURES OF CARCINOGENS

The subcommittee believes that additivity will usually apply to exposure to carcinogens associated with low risks. However, like the National Research Council Committee on Methods for the In Vivo Toxicity Testing of Complex Mixtures, this subcommittee is aware that large exposures to several carcinogens have been shown to produce synergistic interactions—e.g., in a number of studies cigarette-smoking and exposure to asbestos appear to have combined to produce a greater than additive risk of cancer—although the mechanisms of carcinogenesis (in this case of asbestos and cigarette smoke—itself a highly complex mixture) might be different. The subcommittee assumed that exposures to waterborne toxicants, such as VOCs, at low concentrations generally are associated with low individual component risks, although the empirical evidence for such an assumption is sparse. Given that assumption, it should be possible to estimate the risk associated with exposure to a mixture of carcinogens by adding the calculated carcinogenic responses to the individual components of the mixture. The current methods and models used by EPA to estimate carcinogenic risks in humans on the basis of experimental exposures of animals at high doses are derived from the same assumption. Additional research should be conducted to provide a firmer empirical base for the models used.

OTHER ISSUES

The subcommittee is aware of the possibility of substantially increased risk to some persons associated with even small exposures to chemicals, but it did not address this extensively. Because of the genetic variability of humans or because of earlier sensitizing exposures, what is apparently low-dose exposure for the majority of the population might have serious effects in a small segment of the same population. Providing advice on how to factor these complexities into risk assessment was beyond this subcommittee's charge.

1

Introduction

In light of the task before the U.S. Environmental Protection Agency (EPA) of regulating exposure to toxic chemicals, its Office of Drinking Water asked the Safe Drinking Water Committee of the National Research Council's Board on Environmental Studies and Toxicology to hold a small workshop, with each participant addressing some aspect of the methodology for assessing the risk associated with exposure to mixtures of chemicals found in drinking water. This report is the product of that workshop, held in October 1987 in Washington, D.C., and of the deliberations of the Subcommittee on Mixtures in a followup meeting.

This chapter briefly describes the background of the workshop, defines concepts and terms, and suggests ways of grouping chemicals for estimating their combined risk.

BACKGROUND OF THE STUDY

More than 6 million chemicals have been listed and given identifying numbers by the Chemical Abstracts Service of the American Chemical Society. Most have not been adequately tested for toxicity (NRC, 1984) either individually or in combinations. Some 67,000 of them are registered with federal regulatory agencies for use as industrial chemicals; as pesticide, food, drug, and cosmetic ingredients; or for other commercial purposes. Industrial discharges or nonpoint discharges, such as runoff from hazardous-waste sites or agricultural application, might cause many of those chemicals to appear in surface water or groundwater and hence in drinking water. One consumer advocacy group (Center for Study of Responsive Law) has compiled a list of some 2,000 contaminants in drinking water as detected in surveys conducted since 1974 (Duff Conacher and Associates, 1988). Contaminants and their concentrations vary with site, time, and temperature, and many (including the by-products of disinfection) have not been characterized or even identified (NRC, 1987). Because of the variability of drinking water composition and because the relatively low concentrations of the chemical contaminants in water would require large lifetime studies to reveal long-term effects, regulatory authorities and theoretical scientists have attempted to model the effects of mixtures by using the results of tests of the individual components of the mixtures, often for shorter periods at higher doses (Bingham and Morris, 1988; NRC, 1988). The National Toxicology Program has initiated short-term and subchronic studies of a mixture of 25 groundwater contaminants at concentrations actually encountered (Yang and Rauckman, 1987); however, almost nothing is known about how chemicals interact when they are ingested by humans as mixtures or with substances from other sources, including medications. The potential for interactions that could have adverse health consequences must be considered in any assessment of the quality of drinking water. A recent study showed a statistically significant association between the ingestion of chlorinated surface water and human bladder cancer (Cantor et al., 1987). Although the specific components responsible for that association remain unidentified, the by-products of chlorine disinfection are currently the prime suspects.

CONCEPTS AND DEFINITIONS

Exposure to two or more chemicals simultaneously can produce interactions that qualitatively or quantitatively differ from biologic responses that would be predicted from the actions of the individual chemicals separately (Murphy, 1980; NRC, 1980, 1988). When the response is greater than that predicted on the basis of adding the separate responses, the interaction is said to be synergistic. When the response is less than that predicted on the basis of additivity, the interaction is said to be antagonistic.

There is a contrast between dose additivity and response additivity that needs to be addressed. If two (apparently different) toxic materials lead to the same type and severity of toxic effect, they might be considered as one material in the effect they produce. When there is no threshold and the dose-response curve is essentially linear (at least across some modest range of doses), a response-additive model is reasonable. Under such circumstances, a response-additivity model and a dose-additivity model will give the same answer for the same range of doses.

However, if some minimum dose must be reached before toxicity is manifest (i.e., if there is a threshold), then the response-additivity model can be misleading. At subthreshold doses, each of the materials in a mixture will produce zero response separately. A response-additivity model would predict that the sum of any number of zero responses will be a zero response. However, if the sum of the doses is above the threshold, then a response can occur and lead to an appearance of synergism—i.e., greater than response additivity. For a more technical definition and more detailed models of synergism, see Elashoff et al. (1987) and Fears et al. (1988, 1989).

For computational purposes for distinguishing between possible additivity and multiplicativity of responses, a dose that leads to a relative risk of 1.01 or less is called a "low" dose. That does not, in any way, imply that such a dose is "acceptable" or "safe." For example, if the overall age-adjusted mortality from cancer in 1 year in the United States is 200×10^{-5}, a dose leading to a 1% increase would imply an excess mortality of 2×10^{-5}, which is generally considered to be unacceptable, although the relative risk, $(202 \times 10^{-5})/(200 \times 10^{-5})$, equals 1.01, an increase (as defined) of "only" 1%. Exposure to two materials, each at such a dose, would, if results were strictly response-additive, produce a relative risk of 1.02 (i.e., $1 + 0.01 + 0.01$). If results were multiplicative, the relative risk would be $(1.01)(1.01) = 1.0201$—implying an excess over an additive risk that is extremely unlikely ever to be measured or even measurable.

Synergistic interactions between chemicals have been suspected of causing health effects in humans that could not be predicted by simply adding the expected effects of the component chemicals. One such case was the incident of mass organophosphorus insecticide poisoning among field workers in Pakistan in 1976 (Baker et al., 1978); two of the pesticide formulations contained contaminants, which could well have increased the toxicity attributed to the designated active pesticidal ingredient, malathion, by inhibiting its detoxification (see Chapter 4).

The present approach for regulating organic chemicals in drinking water (EPA, 1987) is to establish a maximum contaminant level goal (MCLG) and a maximum contaminant level (MCL) for each organic compound, except that trihalomethanes (THMs) as a class are regulated by a single MCL (EPA, 1979). That exception is based on the "potential" carcinogenicity of chloroform in humans and the similarities of chloroform to less-studied THMs; chloroform is assumed to be representative of a class (the THMs) that is ubiquitous in treated drinking water in the United States and whose members' concentrations can be reduced simultaneously. All other EPA standards are established after the toxicologic data, treatment capabilities, and occurrence data are interpreted and evaluated for each chemical. The single-chemical approach is scientifically appealing, but it could pose major problems, because it ignores both the possibility of interaction and the presence of many unidentified chemicals (NRC, 1987) in treated drinking water.

GROUPING CHEMICALS FOR ESTIMATION OF COMBINED RISK

Methods for assessing health risks associated with mixtures have not changed substantially in recent years (EPA, 1985, 1986; Murphy, 1980; NRC, 1980, 1988). Lack of toxicologic information and the complexity of a mixture can impede and complicate the application of any method, and considerable reliance is still placed on knowledge of the toxicity of individual chemicals in approaching the regulation of mixtures. Because of the substantial backlog in testing and regulation, regulatory agencies need to explore carefully some ways to group chemicals so as to facilitate their control in the absence of complete toxicologic information. Grouping could be used to establish priorities for testing, to formulate rules for testing, or to develop standards for allowable concentrations of contaminants in drinking water.

The subcommittee did not consider either the complete universe of classifications that might be devised or the regulatory consequences of implementing standards for classification. It did, however, review options and suggested the following four types of grouping for consideration by EPA:

1. Contaminants can be grouped on the basis of their being carcinogenic. According to the currently preferred dose-extrapolation models (EPA, 1986; NRC, 1987), the risk of one end point associated with exposure to a mixture of carcinogens at low concentrations can be theoretically approximated as the sum of risks associated with the individual carcinogens; i.e., additivity of response or risk is usually assumed for carcinogens associated with relative risks of less than 1.01. However, the subcommittee recognizes that this assumption of low-dose additivity of response does not have much empirical foundation. Rather, it rests on theoretical considerations and observations from limited epidemiologic studies, and it might not apply for carcinogens at doses yielding high relative risks or when alternative dose-extrapolation models are considered. For exposures at higher concentrations, synergistic interactions appear to occur in humans exposed to combinations of several kinds of agents—such as cigarette smoke, asbestos, and alcoholic beverages (NRC, 1988). The assumption of low-dose additivity needs to be carefully assessed in future research.

2. Systemic contaminants that have similar toxic end points, such as those resulting in specific organ toxicity or peripheral nerve damage, can be grouped and treated as having additive effects under most conditions. A general description of this approach is given in Chapter 3, and the anticholinesterases, which have similar toxic consequences, are examined in detail as a biologically based class in Chapter 5. Materials that are assumed to have thresholds for response require special attention to the biologic mechanisms leading to a toxic response. As indicated earlier, where the mechanisms of toxicity of two or more toxicants are the same, combining below-threshold doses (i.e., doses that produce a zero response) could lead to an above-threshold dose,

(i.e., a dose that produces a response). Thus, dose additivity should be considered for materials that yield the same toxic end point. Furthermore, in some cases in which toxicity information is limited and exposure concentrations are high, an uncertainty factor could be applied to accommodate the possibility of synergism; this is discussed in Chapter 3.

3. Dissimilar chemical compounds might be physiologically transformed into similar metabolites with similar reactivity and stability. For the purposes of assessing combined risk, chemicals can be grouped on the basis of this kind of similarity. A caution to keep in mind is that materials that appear to have similar metabolites might nonetheless at times have different toxic end points (see Chapter 5).

4. Contaminants can be grouped according to structural similarity, which might imply similar biologic responses.

In grouping chemical mixtures by whatever method, a "toxic-equivalence" approach can be considered—assigning numerical potency values to individual mixture components that are representatives of specific classes, estimating potencies of other class members that are present relative to those of the appropriate representative chemicals, and then summing the products of the relative potencies and concentrations of all the chemicals present across all end points. Risks associated with exposure to polycyclic aromatic hydrocarbons (Clement Associates, 1988) and the chlorinated dibenzo-*p*-dioxins and dibenzofurans (Bellin and Barnes, 1987) have been estimated by this method. Again, the concept of dose additivity is inherent in the consideration of toxic equivalents and relative potency; this approach implies that one material is operationally a dilution (in effect) of the other material.

Volatile halogenated hydrocarbons—including carbon tetrachloride, tetrachloroethylene, trichloroethylene, and 1,2-dichloroethane—are frequently found in drinking water, and several could be placed into more than one of the above groups (e.g., into the group of carcinogens or into a group of chlorinated compounds). These substances have similar physical and chemical properties, they are metabolized in the liver, and similar methods are used in treating drinking water to reduce their concentrations. The highest recommended concentrations for these compounds are all set within a rather narrow range (2–5 µg/liter), because their parallel toxicologic properties are similar and it is feasible to control them together (EPA, 1987). The chemicals could be evaluated together with an additivity formula or with a single standard for a mixture of them. Combining the carcinogenic potentials of such chemicals is discussed in Chapter 7.

STRUCTURE OF THE REPORT

Chapter 2 explains the importance of pharmacokinetics to an estimation of the health risks associated with multiple-chemical exposures. Appendix

A gives examples of the use of physiologically based pharmacokinetics to determine the risk to humans of exposure to volatile synthetic organic compounds by extrapolating data from inhalation studies performed on laboratory animals. The current EPA approach for assessing health risks related to noncarcinogens is reviewed in Chapter 3, which suggests modifications of the approach. A simple illustrative mathematical model that supports some of the modifications is given in Appendix B. Appendix C discusses dose additivity and response additivity. Chapter 4 considers issues of exposure, with emphasis on the organophosphates, carbamates, and volatile organic compounds. Chapter 5 reviews the biologic mechanisms of and interactions among anticholinesterases. Chapter 6 shows how the assumptions inherent in EPA's risk assessment methods for carcinogens can be used to combine the estimated risks associated with individual carcinogenic components in a mixture. Although the workshop led to an affirmation of current methods for the risk assessment of mixtures in drinking water, attempts at developing a firmer empirical base and reevaluation should continue. Chapter 7 recommends research to facilitate further improvement.

Because there are so many chemicals and so many possible mixtures to which humans could be exposed, the absence of toxicity data might result in human exposure to chemicals or mixtures that are not being studied by regulatory agencies. More research is needed on the scientific basis for grouping chemicals for testing and regulation. Specific research proposals are given in Chapters 2–6 and summarized in Chapter 7.

REFERENCES

Baker, E. L., Jr., M. Warren, M. Zack, R. D. Dobbin, J. W. Miles, S. Miller, L. Alderman, and W. R. Teeters. 1978. Epidemic malathion poisoning in Pakistan malaria workers. Lancet 1(8054):31–34.

Bellin, J. S., and D. G. Barnes. 1987. Interim Procedures for Estimating Risk Associated with Exposures to Mixtures of Chlorinated Dibenzo-p-dioxins and Dibenzofuran (CDDs and CDF). U.S. Environmental Protection Agency Report No. EPA/625/3–87/012. Washington, D.C.: Risk Assessment Forum, U.S. Environmental Protection Agency. 27 pp. + appendixes.

Bingham, E., and S. Morris. 1988. Complex mixtures and multiple agent interactions: The issues and their significance. Fund. Appl. Toxicol. 10:549–552.

Cantor, K. P., R. Hoover, P. Hartge, T. J. Mason, D. T. Silverman, R. Altman, D. F. Austin, M. A. Child, C. R. Key, L. D. Marrett, M. H. Myers, A. S. Narayana, L. I. Levin, J. W. Sullivan, G. M. Swanson, D. B. Thomas, and D. W. West. 1987. Bladder cancer, drinking water source, and tap water consumption: A case-control study. J. Natl. Cancer Inst. 79:1269–1279.

Clement Associates. 1988. Comparative Potency Approach for Estimating the Cancer Risk Associated With Exposure to Mixtures of Polycyclic Aromatic Hydrocarbons. Interim final report to the U.S. Environmental Protection Agency, Office of Research and Development, April 1, 1988. Contract No. 68–02–4403. Vienna, Va.: Clement Associates, Inc.

Duff Conacher and Associates. 1988. Troubled Waters on Tap: Organic Chemicals in Public Drinking Water Systems and the Failure of Regulation. Washington, D.C.: Center for Study of Responsive Law. 120 pp.

Elashoff, R. M., T. R. Fears, and M. A. Schneiderman. 1987. The statistical analysis of a carcinogen mixture experiment. I. Liver carcinogens. J. Natl. Cancer Inst. 79:509–526.

EPA (U.S. Environmental Protection Agency). 1979. National interim primary drinking water regulations; control of trihalomethanes in drinking water. Final Rule. Fed. Regist. 44(231):68624.

EPA (U.S. Environmental Protection Agency). 1985. Proposed guidelines for the health risk assessment of chemical mixtures. Fed. Regist. 50(6):1170–1176.

EPA (U.S. Environmental Protection Agency). 1986. Guidelines for the health risk assessment of chemical mixtures. [FRL-2984–2.] Fed. Regist. 51(185)34014–34025.

EPA (U.S. Environmental Protection Agency). 1987. National primary drinking water regulations; Synthetic organic chemicals; Monitoring for unregulated contaminants. Fed. Regist. 52(130):25690–25717.

Fears, T. R., R. M. Elashoff, and M. A. Schneiderman. 1988. The statistical analysis of a carcinogen mixture experiment. II. Carcinogens with different target organs, N-methyl-N-nitro-N-nitrosoguanidine, N-butyl-N-(4–hydroxybutyl)nitrosamine, dipentylnitrosamine and nitrilotriacetic acid. Toxicol. Indus. Health 4:221–255.

Fears, T. R., R. M. Elashoff, and M. A. Schneiderman. 1989. The statistical analysis of a carcinogen mixture experiment. III. Carcinogens with different target systems, aflatoxin B1, N-butyl-N-(4–hydroxybutyl)nitrosamine, lead acetate, and thiouracil. Toxicol. Ind. Health 5:1–23.

Murphy, S. D. 1980. Assessment of the potential for toxic interactions among environmental pollutants. Pp. 277–294 in The Principles and Methods in Modern Toxicology, C. L. Galli, S. D. Murphy, and R. Paoletti, eds. New York: Elsevier/North-Holland Biomedical Press.

NRC (National Research Council). 1980. Principles of Toxicological Interaction Associated with Multichemical Exposures. Washington, D.C.: National Academy Press. 213 pp.

NRC (National Research Council). 1984. Toxicity Testing: Strategies to Determine Needs and Priorities. Washington, D.C.: National Academy Press. 382 pp.

NRC (National Research Council). 1987. Drinking Water and Health. Vol. 7. Disinfectants and Disinfectant By-Products. Washington, D.C.: National Academy Press. 207 pp.

NRC (National Research Council). 1988. Complex Mixtures: Methods for In Vivo Toxicity Testing. Washington, D.C.: National Academy Press. 227 pp.

Yang, R. S. H, and E. J. Rauckman. 1987. Toxicologicial studies of chemical mixtures of environmental concern at the National Toxicology Program: Health effects of groundwater contaminants. Toxicology 47:15–34.

2

Pharmacokinetics and the Risk Assessment of Drinking Water Contaminants

This chapter discusses the uses of physiologically based pharmacokinetics in risk assessment. It considers the extrapolation of data from inhalation studies for assessing the risk associated with ingesting drinking water contaminants. Finally, it discusses the pharmacokinetics related to interactions of multiple chemicals found in drinking water.

Some of the more uncertain aspects of risk assessment are related to the extrapolation of data from animals to humans, from one route of exposure to another, from high doses to low doses, and, for carcinogens, from one target organ to another. The reason that data must be extrapolated is that some important kinds of experimental work are impossible, impractical, or unethical. For example, human experiments involving carcinogens are unethical, and animal experiments to study infrequent or very small responses are impractical. To extrapolate among species, doses, routes, and exposure times, one must make assumptions. The assumptions are usually based on scientific facts, informed guesses, or intuition.

The use of pharmacokinetics in the risk assessment of single-chemical exposure has been promoted by some scientists for many years (Andersen et al., 1987a; Clewell and Andersen, 1985; Dedrick, 1985; Gehring et al., 1978; Hoel, 1985; Hoel et al., 1983; Lutz and Dedrick, 1985; NRC, 1986, 1987). Until recently, however, the examples available in the literature were based on classical or conventional compartmental pharmacokinetic studies (Curry, 1980; Gibaldi and Perrier, 1982; O'Flaherty, 1981; Renwick, 1982; Wagner, 1975; WHO, 1986). For applications to toxicology, the classical pharmacokinetic studies were intrinsically weak in interspecies extrapolation, because they were largely mathematical manipulations of experimental data with limited incorporation of physiologic responses or anatomic entities into the model. The current approach in pharmacokinetics includes both physiologically based pharmacokinetics and computer modeling.

The concepts of physiologically based pharmacokinetics and animal "scaleup" (a term adapted from chemical engineering to express the "allometric" extrapolation from one animal species to another or from laboratory animals to humans) originated in the 1920s. They were expanded in the late 1960s and early 1970s with the development of cancer chemotherapy in laboratory animals by investigators experienced in chemical engineering process design and control (Bischoff and Brown, 1966; Bischoff et al., 1970, 1971; Dedrick, 1973a,b; Dedrick et al., 1970). The scaleup from a mouse to a human, like the scaleup from a chemical engineering process in the laboratory to a full-scale chemical plant, is governed by both physical and chemical processes. In mammals, the physical processes (i.e., mass balances, thermodynamics, transport, and flow) often vary predictably among species, whereas chemical processes, such as metabolic reactions, can vary unpredictably. The physical and chemical processes interact in such a way that the pharmacokinetics of a given compound in one species might be predicted from observations of its pharmacokinetics in another species, given the appropriate background information (Dedrick, 1973a,b), but potential problems are numerous, and direct validation of a pharmacokinetic model is generally not possible.

PHYSIOLOGICALLY BASED PHARMACOKINETICS

A physiologically based pharmacokinetic model uses basic physiologic and biochemical data to describe the distribution and disposition of xenobiotic compounds in the body at any given time (NRC, 1987). MacNaughton et al. (1983) and Andersen (1987) summarized the approach in a flowchart (Figure 2-1). Information is categorized into three types: (1) physiologic constants, including body size, organ and tissue volumes, blood flow, and ventilation rates; (2) biochemical constants, including metabolic rates and partition coefficients for blood, tissues, and air; and (3) mechanistic factors, such as target tissues and metabolic pathways.

For the most-studied compounds, the biochemical constants, such as K_m (the affinity constant of an enzyme for a substrate) and V_{max} (the maximal velocity of a chemical reaction), are often available from the literature. Physiologic constants, such as organ volumes and blood flow rates for common laboratory animals, are also available. Therefore, for well-studied chemicals, a dynamic model can be formulated to describe distribution and disposition with little or no further laboratory work. A model can be graphically illustrated, as shown in Figure 2-2, and mathematically represented by many (sometimes 20 or more) simultaneous differential equations to express mass

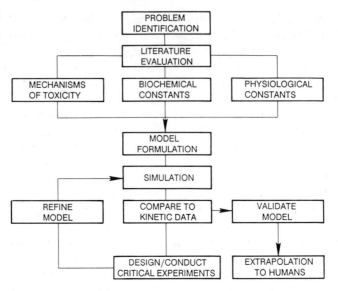

FIGURE 2-1   Flowchart illustrating processes involved in physiologically based pharmacokinetics. From Andersen, 1987.

balance. These cannot in general be solved explicitly, but computer simulations can estimate changes in end points over time, as well as steady states (such as blood concentrations of the parent compound and liver concentrations of a reactive metabolite), and similar information can be extrapolated for different species at lower or higher doses, via different routes of exposure, or both.

The simulated data can then be compared with the experimental kinetic data found in the literature. As Andersen et al. (1987a) emphasized, the validation of a physiologically based pharmacokinetic model is not an exercise in curve-fitting, and experimental data for validation should be obtained after the a priori prediction. A completely validated model is not easily obtainable, but agreement indicates that simulation results are appropriate, compared with experimental reality. If the model is adequately validated, it can be used to extrapolate, directly or by computer simulation, to other animal species (for further validation) or to humans. Lack of agreement means that the model is deficient and that the investigator needs more scientific information, which can be obtained from focused experiments designed to help to refine the model. The refinement process can be repeated for further improvement.

Physiologically based pharmacokinetic models use a large body of physiologic and physicochemical data that are not chemical-specific; they allow interspecies extrapolation with more confidence; they can be used to predict a priori the pharmacokinetic behavior of some chemicals from sparse data; their compartments correspond to anatomic entities, so organ- or tissue-specific biochemical interactions can be incorporated (Dedrick, 1973a,b); and they are more complex and versatile than compartmental pharmacokinetic models. In the past, the application of physiologically based pharmacokinetics was limited to a few investigators because of the complexity of the mathematics involved, the large numbers of parameters in the models, and the requirement for simultaneous solution of many differential equations. In recent years, advances in computer science and readily available software for personal computers have overcome most of the computational limitations.

The model illustrated in Figure 2-2 reflects basic mammalian physiology and anatomy with compartmental entities, such as the liver and kidney, connected by the circulatory system. In this specific model, the exposure route of interest is inhalation, with intake and exhalation vapor concentrations indicated. However, oral or cutaneous exposures can be added to the gastrointestinal tract compartment or general venous circulation. Some tissues (e.g., viscera and brain in Figure 2-2) can be lumped together, when there is no reason to believe that they are kinetically or mechanistically distinct

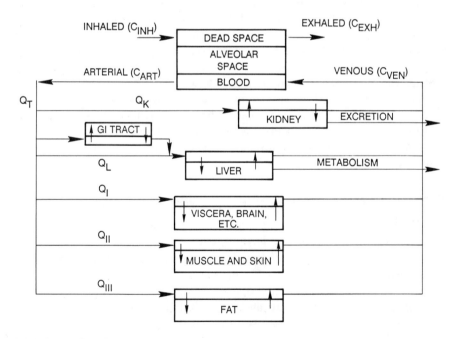

FIGURE 2-2 Graphic representation of a physiologically based pharmacokinetic model. C, concentration of chemical of interest; Q, flow rate; direction of arrow indicates direction of movement of chemical of interest. From Clewell and Andersen, 1985, with permission.

enough to warrant separate compartments. The arrows in each compartment depict the partition of the chemical between blood and organ tissues. Some models might have as few as two compartments; most would have more. They can be flow-limited, like the example given here, or membrane-limited, as suggested by Himmelstein and Lutz (1979). The kinetic constants and model parameters used in pharmacokinetic modeling can be illustrated best with an actual example, such as methylene chloride, as in Table 2-1 (Andersen et al., 1987a).

The kinetic constants and model parameters listed in Table 2-1 for humans and three laboratory species were mainly the results of direct use, estimation or deduction based on scientific reasoning, or extrapolation from published information; the investigators had relatively few new laboratory data before pharmacokinetic modeling. However, the agreement is excellent between the model predictions (a priori) and the experimental kinetic data, which were obtained from three laboratories (Andersen et al., 1984; Angelo et al., 1984; R. H. Reitz, Toxicology Research Laboratory, Dow Chemical Co., Midland, Michigan, personal communication, 1988) through two routes of administration (intravenous and inhalation) in four species (B6C3F1 mice, Syrian golden hamsters, Fischer 344 rats, and humans).

The computer modeling of physiologically based pharmacokinetics is evolving. It is a powerful tool, and the modeling needs to incorporate some form of uncertainty analysis, which is not usually done now. With so many parameters involved, there is no clear relationship between the effects of parameter errors and predicted errors; nor are there clear tests of adequacy of the fit of the model when the parameter estimates have multiple sources. Sensitivity analyses of parameters involved will be important for the improved understanding of physiologically based pharmacokinetic modeling, the design of research to improve models, and the interpretation and application of the results. Cohn (1987) has published a critical discussion on this issue with a specific example.

EXTRAPOLATION BETWEEN INHALATION AND DRINKING WATER ROUTES

Pharmacokinetic studies of chemicals in drinking water or feed are often difficult to conduct. Water and food intakes in rodents, the most commonly used laboratory animals, are episodic and erratic. In addition, rodents are nocturnal, and most of their drinking and eating occur at night. Those factors make sampling of body fluids and tissues difficult. It is not only a problem of time of day (e.g., multiple sampling in the middle of the night, when animal facilities are in the dark part of the cycle), but also a problem of informed guesses about the time of peak blood concentrations (sampling

TABLE 2-1 Kinetic Constants and Model Parameters Used in the Physiologically Based Pharmacokinetic Model for Methylene Chloride[a]

	B6C3F1 Mice[b]	F344 Rats[c]	Hamsters[c]	Humans
Weights				
Body (kg)	0.0345	0.233	0.140	70.0
Lung (g)	0.410	2.72	1.64	772.0
Percentage of body weight				
Liver	4.0	4.0	4.0	3.14
Rapidly perfused tissue	5.0	5.0	5.0	3.71
Slowly perfused tissue	78.0	75.0	75.0	62.1
Fat	4.0	7.0	7.0	23.1
Flows (liters/hour)				
Alveolar ventilation	2.32	5.10	3.50	348.0
Cardiac output	2.32	5.10	3.50	348.0
Percentage of cardiac output				
Liver	0.24	0.24	0.20	0.24
Rapidly perfused tissue	0.52	0.52	0.56	0.52
Slowly perfused tissue	0.19	0.19	0.19	0.19
Fat	0.05	0.05	0.05	0.05
Partition coefficients				
Blood/air	8.29	19.4	22.5	9.7
Liver/blood	1.71	0.732	0.840	1.46
Lung/blood	1.71	0.732	0.840	1.46
Rapidly perfused tissue blood	1.71	0.732	0.840	1.46
Slowly perfused tissue blood	0.960	0.408	1.196	0.82
Fat/blood	14.5	6.19	6.00	12.4
Metabolic constants				
V_{max} (mg/hour)	1.054	1.50	2.047	118.9
K_m (mg/liter)	0.396	0.771	0.649	0.580
KF (hour^{-1})	4.017	2.21	1.513	0.53
A1[d]	0.416	0.136	0.0638	0.00143
A2[d]	0.137	0.0558	0.0774	0.0473

[a]From Andersen et al., 1987a, with permission. Copyright 1987 by Academic Press.
[b]Parameters correspond to average body weight of B6C3F1 mice in NTP bioassay (NTP, 1985).
[c]Parameters correspond to average body weight in gas-uptake studies.
[d]A1 = ratio of MFO (mixed-function oxidase) activity in lung to MFO activity in liver. A2 = ratio of GST (glutathione S-transferase) activity in lung to GST activity in liver.

points at critical stages can be missed). In addition, many drinking water contaminants are volatile, lipophilic, organic compounds and are likely to be unstable in drinking water or feed formulations. If radioactive compounds are to be used in the study, the potential contamination problems with respect to the animals, equipment, and facility are difficult to handle. Even in the absence of those problems, it is hard to interpret, for example, a blood concentration-time curve with peaks of different heights and shapes at irregular intervals.

A National Research Council report (NRC, 1986) suggested an approach to overcome the above problems. It used pharmacokinetic data on volatile organic chemicals (VOCs) from inhalation studies for the risk assessment of exposure to these compounds through the ingestion of drinking water. A brief summary of an example using trichloroethylene (TCE) is given in Appendix A; a more detailed discussion appears in the report just mentioned (NRC, 1986).

PHARMACOKINETICS INVOLVING INTERACTIONS

Physiologically based pharmacokinetic modeling of toxic interactions is a new field, and the only published studies are limited to binary mixtures (Andersen et al., 1987b; Clewell and Andersen, 1985). Andersen et al. (1987b) illustrated the use of physiologically based pharmacokinetic modeling of the metabolic interactions between TCE and 1,1-dichloroethylene (1,1-DCE). A physiologic model was constructed for each of the two compounds individually, and the two models were linked via the mass-balance equation for the liver compartment that had been generalized to account for various mechanisms of interaction between the two compounds. The generalized scheme was used to account for inhibitory interactions—including provisions for competitive, noncompetitive, and uncompetitive mechanisms—as well as for substrate inhibition. The correspondence between predicted and observed kinetics was excellent, if it could be assumed that the inhibition was purely competitive and if 1,1-DCE was considered to be a slightly better substrate for microsomal oxidation than TCE in the model. Figure 3-3 shows two uptake curves for 1,1-DCE in gas-uptake experiments; one is for exposure to 1,1-DCE alone at 500 ppm, and the other is for exposure to a vapor mixture of 1,1-DCE at 500 ppm and TCE at 2,000 ppm. The disappearance of 1,1-DCE (as a result of metabolism) was markedly retarded when coexposure with TCE was carried out. When the scientific hypothesis was based on known biology, the a priori prediction and the experimental kinetic data agreed very well (Figure 2-3).

PHARMACOKINETICS AND TOXIC MECHANISMS OF MULTIPLE CHEMICAL EXPOSURE

Recent discussion of the role of pharmacokinetics in the study of complex mixtures (NRC, 1988) has emphasized that little is known about the joint pharmacokinetics of two or more chemicals. Generation and examination of such data have been suggested (Yang, 1987a,b), but the application of phar-

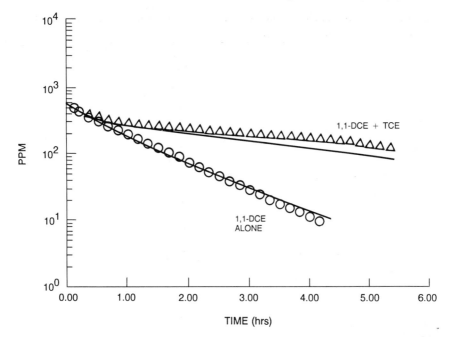

FIGURE 2-3   Two uptake curves for 1,1-dichloroethylene (1,1-DCE) from experimental gas-uptake studies (circles and triangles) and from physiologically based pharmacokinetic models (smooth curves), assuming strictly competitive interactions between two chloroethylenes. Lower curve, exposure to 1,1-DCE alone at 500 ppm. Upper curve, exposure to 1,1-DCE at 500 ppm and trichloroethylene (TCE) at 2,000 ppm. From Andersen et al., 1987b, with permission. Copyright 1987 by Academic Press.

macokinetics to the risk assessment of multiple chemical exposures through contaminated drinking water remains difficult and subject to large uncertainties. Several toxicologic studies (Chu et al., 1981; Côté et al., 1985; Webster et al., 1985) have dealt with the health effects of exposures to multiple chemicals at low doses, including a carcinogenicity study. Thus, some toxicologic information can be used in the risk assessment of multiple chemicals, although the mixtures in those studies are of only selected classes of chemicals (e.g., halogenated volatile organic chemicals, inorganic chemicals, and pesticides). A mixture of 25 groundwater contaminants (Table 2-2), selected on the basis of EPA surveys of groundwater in and around hazardous-waste disposal sites, is being evaluated toxicologically by the National Toxicology Program (Yang and Rauckman, 1987), but the results of relatively long-term studies, are not yet available. Methods for risk assessment of mixtures of chemicals in drinking water are still based largely on speculation, and no quick relief is in sight.

Although a small fraction of the U.S. population living close to hazardous-

TABLE 2-2   Groundwater Contaminants Selected for Study as a Mixture by the National Toxicology Program[a]

Chemical	Concentrations in Groundwater Samples, ppm	
	Average	Highest
Acetone	6.9	250
Arochlor 1260	0.21	2.9
Arsenic	30.6	3,670
Benzene	5.0	1,200
Cadmium	0.85	225
Carbon tetrachloride	0.54	20
Chlorobenzene	0.1	13
Chloroform	1.46	220
Chromium	0.69	188
1,1-Dichloroethane	0.31	56.1
1,2-Dichloroethane	6.33	440
1,1-Dichloroethylene (1,1-DCE)	0.24	38.0
1,2-trans-Dichloroethylene	0.73	75.2
Di-(2-ethylhexyl)phthalate (DEHP)	0.13	5.8
Ethylbenzene	0.65	25
Lead	37.0	31,000
Mercury	0.34	50
Methylene chloride	11.2	7,800
Nickel	0.5	95.2
Phenol	34.0	7,713
Tetrachloroethylene	9.68	21,570
Toluene	5.18	1,100
1,1,1-Trichloroethane	1.25	618
Trichloroethylene (TCE)	3.82	790
Xylenes	4.07	150

[a]Condensed from Yang and Rauckman, 1987, with permission; analytic survey of groundwater samples in and around 180 hazardous-waste sites covering all 10 EPA regions. Survey conducted for EPA by Lockheed Engineering and Management Co.

waste disposal sites might be consuming groundwater containing one or more of the chemicals listed at near the average concentrations shown, the concentrations of contaminants in public drinking water supplies used by most Americans (see Table 4-1) are much lower than the averages listed in Table 2-2. Consideration of the hypothetical mixture of 25 chemicals (Table 2-2—a worst case)—can yield insight into the possible pharmacokinetic and toxic consequences of consuming drinking water that contains multiple contaminants.

On the basis of the toxicity of the individual chemicals, it is probably safe to suggest that none of the 25 (Table 2-2) taken singly (for example, in an 8-ounce glass of water) at the average concentration found in drinking water surveys would approach the saturation kinetic level unless a genetic variation has deprived a person of a pathway. However, under the conditions of acute exposure at very high concentrations (e.g., the highest listed in Table 2-2, or even higher) or repeated or chronic exposure at lower concentrations (e.g., the average in Table 2-2), the situation could be quite different. Given the usual dose-response relationships, each organic chemical in a sample of contaminated drinking water probably has little toxic consequence at low concentrations. Metals, however, tend to accumulate in the body and might therefore pose a long-term health threat. What about toxic interactions under those circumstances? For a mixture containing chemicals in the average amounts found in the published surveys, like the one represented in Table 2-2, it is not clear what toxicity to expect or how to predict it. We know too little for informed speculation about the synergistic effects of the components of such a mixture on toxic end points, such as immunotoxicity, or on such mechanisms as the promotion stage of carcinogenesis. Recent preliminary findings of the National Toxicology Program (Germolec et al., in press) suggested that a mixture of 25 groundwater contaminants, at concentrations close to the averages listed in Table 2-2, is associated with mild but definite immunosuppression in B6C3F1 mice. Those findings merit further examination and suggest that there might be exceptions to the concept of simple response additivity in mixtures of chemicals, or even that the concept is quite broadly wrong. In the absence of adequate information, and to anticipate possible synergism, it might be prudent to incorporate an uncertainty factor in the risk assessment of mixtures of chemicals in drinking water. The development of such an uncertainty factor is considered in more detail in Chapter 3.

CONCLUSIONS AND RECOMMENDED RESEARCH

Physiologically based pharmacokinetic models are useful in the risk assessment of contaminants in drinking water when one or possibly two materials are to be considered. Unfortunately, we know little about how pharmacokinetic variables of a single chemical might be affected in multiple-chemical exposures, nor do we understand the pharmocokinetics of multiple chemicals under such exposure scenarios. Improved understanding and modeling of the pharmacokinetics of mixtures should lead to more accurate estimation of the risks associated with exposure to multiple chemicals in drinking water. Development of appropriate pharmacokinetic models for mixtures will require considerable theoretical and experimental work.

The subcommittee recommends the following research:

• Potential pharmacokinetic changes of individual model chemicals (those which seem representative of others similar in structure, mode of action, or toxic end point) under the influence of long-term, low-concentration intake of a mixture of contaminants in drinking water should be investigated.

• Several physiologically based pharmacokinetic models of complex chemical mixtures simulating contaminated drinking water should be developed and subjected to rigorous validation testing.

• The physiologically based pharmacokinetics of pesticides and some other, relatively nonvolatile chemicals should be studied.

• The frequency of toxic interactions among drinking water contaminants and the threshold concentrations, if any, for such interactions should be investigated.

• A computerized data base on toxic interactions should be built.

REFERENCES

Andersen, M. E. 1987. Tissue dosimetry in risk assessment, or what's the problem here anyway? Pp. 8–23 in Drinking Water and Health, Vol. 8. Pharmacokinetics in Risk Assessment. Washington, D.C.: National Academy Press.

Andersen, M. E., R. L. Archer, H. J. Clewell III, and M. G. MacNaughton. 1984. A physiological model of the intravenous and inhalation pharmacokinetics of three dihalomethanes—CH_2Cl_2, CH_2Br_1, CH_2Br_2—in the rat. Toxicologist 4:443. (Abstract)

Andersen, M. E., H. J. Clewell III, M. L. Gargas, F. A. Smith, and R. H. Reitz. 1987a. Physiologically based pharmacokinetics and risk assessment process for methylene chloride. Toxicol. Appl. Pharmacol. 87:185–205.

Andersen, M. E., M. L. Gargas, H. J. Clewell III, and K. M. Severyn. 1987b. Quantitative evaluation of the metabolic interactions between trichloroethylene and 1,1-dichloroethylene *in vivo* using gas uptake methods. Toxicol. Appl. Pharmacol. 89:149–157.

Angelo, M. J., K. B. Bischoff, A. B. Pritchard, and M. A. Presser. 1984. A physiological model for the pharmacokinetics of methylene chloride in B6C3F1 mice following intravenous administration. J. Pharmacokinet. Biopharm. 12:413–436.

Bischoff, K. B., and R. G. Brown. 1966. Drug distribution in mammals. Chem. Eng. Prog. Symp. Ser. No. 66 62:32–45.

Bischoff, K. B., R. L. Dedrick, and D. S. Zaharko. 1970. Preliminary model for methotrexate pharmacokinetics. J. Pharm. Sci. 59:149–154.

Bischoff, K. B., R. L. Dedrick, D. S. Zaharko, and J. A. Longstreth. 1971. Methotrexate pharmacokinetics. J. Pharm. Sci. 60:1128–1133.

Chu, I., D. C. Villeneuve, G. C. Becking, and R. Lough. 1981. Subchronic study of a mixture of inorganic substances present in the Great Lakes ecosystem in male and female rats. Bull. Environ. Contam. Toxicol. 26:42–45.

Clewell, H. J., III, and M. E. Andersen. 1985. Risk assessment extrapolations and physiological modeling. Toxicol. Ind. Health 1(4):111–131.

Cohn, M. S. 1987. Sensitivity analysis in pharmacokinetic modeling. Pp. 265–272 in Drinking Water and Health, Vol. 8. Pharmacokinetics in Risk Assessment. Washington, D.C.: National Academy Press.

Côté, M. G., G. L. Plaa, V. E. Valli, and D. C. Villeneuve. 1985. Subchronic effects of a mixture of "persistent" chemicals found in the Great Lakes. Bull. Environ. Contam. Toxicol. 34:285–290.

Curry, S. H. 1980. Drug Disposition and Pharmacokinetics: With a Consideration of Pharmacological and Clinical Relationships, 3rd Ed. Oxford: Blackwell Scientific Publications. 330 pp.

Dedrick, R. L. 1973a. Animal scale-up. J. Pharmacokinet. Biopharm. 1:435–461.

Dedrick, R. L. 1973b. Physiological pharmacokinetics. J. Dynamic Syst. Measurement Cont. (Sept.):255–258.

Dedrick, R. L. 1985. Application of model systems in pharmacokinetics. Pp. 187–198 in Risk Quantitation and Regulatory Policy, D. G. Hoel, E. A. Merrill, and F. P. Perera, eds. Cold Spring Harbor, N.Y.: Cold Spring Harbor Laboratory.

Dedrick, R. L., K. B. Bischoff, and D. S. Zaharko. 1970. Interspecies correlation of plasma concentration history of methotrexate (NSC-740). Cancer Chemotherapy Rep. Part I 54:95–101.

Gehring, P. J., P. G. Watanabe, and C. N. Park. 1978. Resolution of dose-response toxicity data for chemicals requiring metabolic activation: Example—vinyl chloride. Toxicol. Appl. Pharmacol. 44:581–591.

Germolec, D. R., R. S. H. Yang, M. P. Ackerman, G. S. Rosenthal, G. A. Boorman, M. Thompson, P. Blair, and M. I. Luster. In press. Toxicology studies of chemical mixtures of 25 groundwater contaminants: (II) immunosuppression in B6C3F$_1$ mice. Fundam. Appl. Toxicol.

Gibaldi, M., and D. Perrier. 1982. Pharmacokinetics, 2nd Ed. New York: Marcel Dekker. 494 pp.

Himmelstein, K. J., and R. J. Lutz. 1979. A review of the applications of physiologically based pharmacokinetic modeling. J. Pharmacokinet. Biopharmacol. 1:127–145.

Hoel, D. G. 1985. Incorporation of pharmacokinetics in low-dose risk estimation. Pp. 205–214 in Biological and Statistical Criteria, D. B. Clayson, D. Krewski, and I. Munro, eds. Toxicologic Risk Assessment, Vol. I. Boca Raton, Fla.: CRC Press.

Hoel, D. G., N. L. Kaplan, and M. W. Anderson. 1983. Implication of nonlinear kinetics on risk estimation in carcinogenesis. Science 219:1032–1037.

Lutz, R. J., and R. L. Dedrick. 1985. Physiological pharmacokinetics: Relevance to human risk assessment. Pp. 129–149 in New Approaches in Toxicity Testing and Their Application in Human Risk Assessment, A. P. Li, ed. New York: Raven Press.

MacNaughton, M. G., M. E. Andersen, and H. J. Clewell III. 1983. Toxicokinetics: An analytical tool for assessing chemical hazards to man. USAF Med. Ser. Digest 39:26–29.

NRC (National Research Council). 1986. Dose-route extrapolations: Using inhalation toxicity data to set drinking water limits. Pp. 168–225 in Drinking Water and Health, Vol. 6. Washington D.C.: National Academy Press.

NRC (National Research Council). 1987. Drinking Water and Health, Vol. 8. Pharmacokinetics in Risk Assessment. Washington, D.C.: National Academy Press. 488 pp.

NRC (National Research Council). 1988. Complex mixtures: Methods for In Vivo Toxicity Testing. Washington, D.C.: National Academy Press. 227 pp.

NTP (National Toxicology Program). 1985. NTP Technical Report on the Toxicology and Carcinogenesis Studies of Dichloromethane in F-344/N Rats and B6C3F1 Mice (Inhalation Studies). NTP-TR-306 (board draft). Research Triangle Park, N.C.: U.S. Department of Health and Human Services.

O'Flaherty, E. O. 1981. Toxicants and Drugs: Kinetics and Dynamics. New York: John Wiley & Sons. 398 pp.

Renwick, A. G. 1982. Pharmacokinetics in Toxicology. Pp. 659–710 in Principles and Methods of Toxicology, A. Wallace Hayes, ed. New York: Raven Press.

Wagner, J. G. 1975. Fundamentals of Clinical Pharmacokinetics. Hamilton, Ill.: Drug Intelligence Publications. 461 pp.

Webster, P. W., C. A. Van Der Heijden, A. Bisschop, G. J. Van Esch, R. C. C. Wegman, and T. De Vries. 1985. Carcinogenicity study in rats with a mixture of eleven volatile halogenated hydrocarbon drinking water contaminants. Sci. Total Environ. 47:427–432.

WHO (World Health Organization). 1986. Principles of Toxicokinetic Studies. Environmental Health Criteria 57. Geneva: World Health Organization. 166 pp.

Yang, R. S. H. 1987a. Acute versus chronic toxicity and toxicological interactions involving pesticides. Pp. 20–36 in Pesticides: Minimizing the Risks, N. N. Ragsdale and R. J. Kuhr, eds. ACS Symposium Series Vol. 336. Washington, D.C.: American Chemical Society.

Yang, R. S. H. 1987b. A Toxicologic View of Pesticides. Chemtech 17:698–703.

Yang, R. S. H., and E. J. Rauckman. 1987. Toxicological studies of chemical mixtures of environmental concern at the National Toxicology Program: Health effects of ground water contaminants. Toxicology 47:15–34.

3

Risk Assessment of Mixtures of Systemic Toxicants in Drinking Water

Even if the resources now devoted to studies of chemical interactions could be multiplied by 1,000 or more, those studies would provide only a small portion of the information required to determine precisely the toxicity of complex mixtures prevailing in the environment. It would still fall to toxicologists and other scientists to integrate the separate pieces of information into a form useful for risk assessment. With the goal of providing adequate but not excessively conservative exposure standards for health protection, this chapter presents various approaches to using incomplete empirical information and scientific judgment to assess the health risks associated with exposure to mixtures in drinking water.

The dominant concern with mixtures, as noted earlier, is unexpected amplification of toxicity arising from combinations of mixture components. The extent of the problem increases with the complexity of the mixture. For example, in a combination of 10 toxicants there are $\binom{10}{2}$, or 45, possible 2-factor interactions;[1] $\binom{10}{3}$, or 120, possible 3-factor interactions; $\binom{10}{4}$, or 210, possible 4-factor interactions; $\binom{10}{5}$, or 252, possible 5-factor interactions; $\binom{10}{6}$, or 210, possible 6-factor interactions; $\binom{10}{7}$, or 120, possible 7-factor interactions; $\binom{10}{8}$, or 45, possible 8-factor interactions; $\binom{10}{9}$, or 10, possible 9-factor interactions; and 1 possible 10-factor interaction. The total number of interactions that must be considered in a study of combinations of 10 substances is:

---

[1] The notation $\binom{10}{2}$ refers to the number of two-substance combinations possible from a group of 10 substances; the general notation, $\binom{m}{c}$, can be rewritten as $m!/c!(m-c)!$, where ! is the symbol for factorial. $m!$ equals $m(m-1)(m-2) \ldots (2)(1)$ i.e., $10! = (10)(9)(8)(7) \ldots (2)(1)$.

121

$$\sum_{i=2}^{10} \binom{10}{i} = 2^{10} - 11 = 1,013. \tag{1}$$

It is difficult to imagine a procedure developed from consideration of the effects of 10 separate agents that adequately accounts for each of the 1,013 possible interactions.

To model interaction requires precise definition of the term "interaction." As described in the 1988 publication by the Committee on Methods for the In Vivo Toxicity Testing of Complex Mixtures (NRC, 1988, p. 100):

Interaction [has been] referred to as deviation from . . . additive behavior expected on the basis of dose-response curves obtained with individual agents. [But] to a biostatistician, the definition includes information on the underlying dose-response model and on units of measure. For ordinary linear models, "interaction" refers to a departure from response additivity. Different measures of response can lead to different conclusions concerning departure from additivity. For nonlinear models, such as log-logistic, log-linear, log-probit, and multistage models, . . . there is additivity for the logarithm of response. Thus, when the word "interaction" is used, one must make certain of the units in which toxicity or dose is expressed, as well as the assumptions about the nature of joint action that was predicted by the model used.

THE PROBLEM OF SYNERGISM

Of greatest concern is the possibility that exposure to mixtures could result in a much more severe toxic response than that expected on the basis of the potencies of the individual components. "Synergism" is here defined as a response greater than additivity.

The mixture problem for drinking water is mainly a problem of relatively low concentrations of individual agents on which human observational data are lacking. Demonstrations of synergism come largely from the laboratory and are usually based on high doses. A limited review of the literature uncovered no cases in which the potency of a particular agent at high doses had been multiplied by more than 100 by combination with another agent. Furthermore, in the extensive literature on drug interactions, novel effects are rarely observed. Instead, if binary combinations act synergistically, they tend to do so through an increase in the effects of one of the agents. Assume, then, that 100 represents the maximal enhancement of toxicity due to the addition of a second agent. That is also the number by which no-observed-adverse-effect levels (NOAELs)—or their advocated replacement, bench-mark doses—are divided to yield acceptable daily intakes (ADIs), or "reference doses" (the EPA term), although the uncertainty factor of 100 is usually designed to protect against uncertainties other than those arising from inter-actions. The number 100 has been said to incorporate a species variation of

10 and a range of individual human susceptibility of 10, though formal documentation of this view is lacking.

THE CONCEPT OF COMMONALITY

One concept that might be useful in combining numerical estimates of toxicity of different substances is that of commonality, formulated by Weiss (1986) in relation to food additives. (Food additives are regulated on the basis of individual ADIs, no explicit account being taken of the possibility that their basic actions might not be mutually independent.) Commonality reflects the extent to which several agents are likely to act on the same target organ and elicit the same toxic response. For instance, several organophosphorus and carbamate chemicals have the commonality of inhibiting acetylcholinesterase. Compounds of the heavy metals mercury and cadmium disrupt renal cellular function and produce renal necrosis.

"Commonality" does not now translate readily into any simple arithmetic manipulations, but Figure 3–1 is an attempt to illustrate the concept. The ADI, or acceptable level (AL), is transformed on the z-axis into total toxic burden or exposure. The x-axis corresponds to the number of chemicals in the mixture to which there is exposure. The y-axis (commonality) represents the overlap in biologic effects of various chemicals. It represents the degree to which the individual constituents of a mixture contribute to a particular process. It recognizes that most agents have broad biologic actions, so their toxic contribution depends on the choice of a specific end point. Fewer agents or a lower commonality would lead to a lower total burden. A higher commonality, assuming a fixed intake as a proportion of the ADI for each agent, would raise the total burden. A commonality factor of less than 1 implies less than full identity of action.

MODIFYING THE HAZARD INDEX

EPA (1986b) has published guidelines for assessing the health risks associated with chemical mixtures. If data on the toxicity of a specific mixture as a whole are unavailable, the risks may be estimated on the basis of what is known about the individual constituents. A hazard index (HI) associated with a mixture of k toxicants may be defined as:

$$HI = \sum_{i=1}^{k} \frac{E_i}{AL_i}, \tag{2}$$

where E_i is the exposure level of the ith toxicant in the mixture and AL_i is the maximum acceptable concentration of the ith toxicant.

When the HI exceeds 1, it evokes the same concerns about the mixture

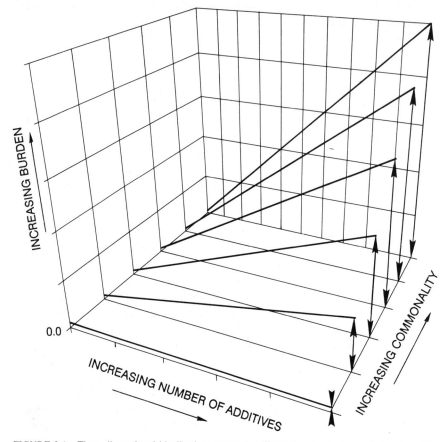

FIGURE 3-1   Three-dimensional idealized representation illustrating increasing toxic burden with the addition of several agents. Lower commonality would reduce total burden; higher commonality would increase it. This illustration applies to only one system or organ at a time. Total burden would be measured in percent of ADI.

as those evoked when any individual AL is exceeded, which is, in fact, a special case, with $k = 1$. This computation of the hazard index is based on the assumption of adding toxically equivalent doses—i.e., the absence of interactions among the components of the mixture. However, such dose-additive models as are implied by the hazard index might not provide the most biologically plausible approach to describing the effects of a complex mixture of toxicants if the compounds do not have the same mode of toxic action. EPA (1986b) has suggested that separate hazard indexes should be developed for separate end points of concern. It further notes that ''dose addition for dissimilar effects does not have strong scientific support, and, if done, should be justified.'' In addition, EPA recognizes that the biologic meaning of the index given above is unknown and that much additional research is required before its utility as a predictor of toxic severity can be validated.

Other Risk-Oriented Approaches

Identification of risk associated with the consumption of chemical mixtures in drinking water will come from many sources: laboratory data, such as those obtained from bioassay-directed fractionation; epidemiologic data, such as those based on similar exposures; and models based on knowledge of the toxicity of individual constituents. When models are the primary source, the assumptions described above and the already available empirical data point to an empirical policy: adoption of the additivity assumption for low-response exposures.

Three strategies for protection against the risks associated with exposure to drinking water contaminants might be defined in this argument. The choice of strategy is a policy issue based on public concern; the data are clearly inadequate to dictate a firm choice.

1. *Least adequate*. Each agent's ADI is determined independently; hence, regulations would consider each agent individually. This alternative is unsatisfactory, because of the low likelihood that each agent can truly act in biologic isolation of every other agent. Furthermore, individually acceptable risks might lose their acceptability when there are 50 or 100 of them.

2. *Usually adequate*. A water quality or hazard index similar to the EPA index would be defined under some assumptions of additivity. Given that ADIs already incorporate uncertainty factors, usually of 100, and given that only one study (Murphy et al., 1959) has reported a 100-fold increase in toxicity through a binary combination (of organophosphates), the simple dose-additive model should be satisfactory under most circumstances. For materials that are assumed to have thresholds for response, special concern must be given to the biologic mechanisms leading to a toxic response. If the mechanisms of toxicity of two or more toxicants are the same, combining below-threshold doses (i.e., doses with zero response) could lead to an above-threshold dose and produce a nonzero response. This concept is inherent in the consideration of toxic equivalents and the idea of relative potency, which implies that one material is (in effect) a dilution of the other material. For these reasons, dose-additivity must be considered for materials that affect the same toxic end point.

3. *Most adequate*. Although most attempts to examine the joint actions of binary combinations find that toxic responses are well described by simple additivity (e.g., Smyth et al., 1969) or that additivity is violated to only a

small degree, maximal protection might require an additional margin of safety beyond that provided by the additivity model. Because of the 100-fold increase observed by Murphy et al. (1959), adjustments of an index based on additivity by an equivalent factor should provide a rather robust safety margin. Such a margin, however, would be excessive in most cases. That is, just as it seems unreasonable to expect totally independent actions of the agents identified in a drinking water source, it seems equally unreasonable to expect complete overlap in toxic targets, so even the additivity alternative already embodies this additional but unrecognized safety factor. Later in this chapter, it is suggested that an uncertainty factor can be considered to compensate for any synergism.

In addition to the EPA hazard index, there are other precedents for combining the toxic manifestations of individual components of a mixture. Threshold limit values (TLVs), which are workplace standards, are combined for simultaneous exposure to more than one agent of the same class of compounds (ACGIH, 1983; NRC, 1980). For example, the joint action of several volatile organic solvents, which induce central nervous system impairment, is estimated by adding the individual ratios of exposure concentration to TLV. For some combinations, such as n-hexane and methylethyl ketone, the assumption of additivity might underestimate toxicity, even though both compounds are organic solvents. The additive model might seem reasonable only for conditions in which the mode of action of the constituents is well understood and greater than additive effects can therefore be excluded on the basis of mechanistic processes. It might be adapted, however, for situations in which drinking water standards, such as MCLs, are based on ADIs derived from chronic studies devoid of mechanistic information.

One way to apply the additive model to extract an overall assessment of drinking water quality is to use the dose-additive procedure adopted by the State of New York to regulate aldicarb and carbofuran. The prescribed limits are 7 and 15 µg/liter, respectively. The actual concentrations at any site are used in the following expression:

$$\frac{(\text{actual aldicarb})}{7} + \frac{(\text{actual carbofuran})}{15} = T. \tag{3}$$

If $T \leq 1$, no action is taken. If $T > 1$, filters are installed, and the cost is charged to the manufacturer whose product is calculated to give the higher ratio. The procedure is based on the assumption that both agents act on identical systems and are totally interchangeable.

A corresponding expression can be based on many more than two agents, as for workplace TLVs. If dose additivity is assumed, as in the case in New York, then the sum of the ratios identified for the hazardous constituents can be required to be no more than 1.0, on the basis of the sum of quotients determined as described above in Equation 3.

The hazard index defined by Equation 2 can be modified to take into account the sensitivity of each toxic end point. As noted earlier, applying the uncertainty factor incorporated in ALs or ADIs is intended to provide a safe (nontoxic) exposure of the most sensitive persons for the toxic manifestation or end point that occurs at the lowest dose if a given material has several manifestations of toxicity. The operating assumption is that an exposure that is safe for the most sensitive end point will be safe for any other toxic manifestation. If sensitivity is defined by the no-observed-adverse-effect level (NOAEL)—the most sensitive end point being defined as the one with the lowest NOAEL—then the relative sensitivities of various end points can be estimated by comparing NOAELs or ratios of NOAELs. That is, sensitivity is taken to be the inverse of the NOAEL, and relative sensitivity is defined as the ratio of the NOAEL for the most sensitive end point to the NOAEL for a specific subsidiary end point. If this relative sensitivity, or ratio of NOAELs, is used as a weighting factor, called W_{ij} (which will always be less than or equal to 1), then the equation for HI can be modified to read:

$$HI_j = \sum_{i=1}^{k} \frac{E_i W_{ij}}{AL_{ij}}, \tag{4}$$

where i is the toxic material and j is some specific end point. This procedure assumes that the AL for each substance is not known for each end point of concern; if it were, $1/AL_{ij}$ could be substituted for W_{ij}/AL_{ij} in the above equation. The HI_j would then be computed by summing HI_j across all materials i. If any of the HI_j values exceeded 1, the sum also must exceed 1, and the actions indicated above for the unmodified HI_j would be taken. This method is sensitive to fractionation by the number of toxicity categories used; thus, it might be most reasonable to sum across all toxic end points:

$$HI = \sum_{j=1}^{l} HI_j, \tag{5}$$

where l is the number of toxic end points considered.

Modifying HI by using the weighting factor W_{ij} takes into account the possibly different toxic spectra of different materials, avoids the lumping together of unrelated toxicities, and still incorporates all reported toxicities into a unified measure.

Grouping Agents with Common Toxicities

With a small number of agents, it might be possible to invoke expert opinion to separate components of toxic mixtures into clusters and to assign an average commonality to each cluster. Or, as an upper limit in the dose-additive model, a commonality of 1.0 could be assumed within each cluster; this would lead to dose additivity. The concept could also be extended by estimating commonalities between clusters defined as chemical classes. For example, clusters as diverse as the heavy metals, organic solvents, and insecticides might all have a given end point, such as peripheral nerve damage. Clustering could be on the basis of mechanistic, toxic, or structural properties. Structural classification, in fact, is the first step by which EPA assesses potential hazards associated with new chemicals, so the process is not unfamiliar for estimating commonalities within a group. With larger numbers or a lack of sufficient data to describe a toxicity profile, such a procedure might prove unwieldy; but commonalities among clusters could still be considered after commonality within a cluster is established. The concept of commonality clearly could be helpful, but more work is needed to develop its application.

Incorporating an Uncertainty Factor for Synergism

The issue of toxic interactions—synergistic or antagonistic—is central in the development of a risk assessment strategy for chemical mixtures in drinking water. Even though the concentrations of contaminants in most sources of drinking water for the general public are likely to be very low, there is insufficient evidence about the toxicity of chemical mixtures after long-term, low-dose exposure to support a definite conclusion that toxic interactions are absent under these conditions. For instance, a combination of chemicals, even at low concentrations, could conceivably act to modify the immune system, thereby compromising natural defense systems. Evidence supporting the existence or absence of such a response process is clearly not available for all relevant mixtures. The argument for the consideration of greater than response additive effects is strengthened by the possibility that water sources in heavily polluted areas (e.g., hazardous-waste sites or point-source accidental spills) contain much higher concentrations of contaminants. Thus, at least a small fraction of the population is sometimes exposed to relatively high (parts per million) concentrations of mixtures of chemicals in drinking water. These types of concerns could lead to approaches that are more protective than those given above.

It is important to address the issue of toxic interactions in the risk assessment of chemical mixtures, but flexibility must be provided to avoid inappropriate regulation. Therefore, the additive approach represented by Equation 5 might be modified by incorporating an additional uncertainty factor (UF), which could be applied to the hazard index defined above to yield:

$$HI = (UF) \sum_{j=1}^{l} HI_j. \tag{6}$$

The UF could vary from 1 to 100, depending on the amount of information available and the concentrations of the contaminants. If a great deal of toxicologic information is available on the individual contaminants, if toxic interactions are not likely (on the basis of the knowledge available), or if the concentrations of the contaminants are "low," the UF might be set at 1 (thus assuming simple additivity). If less is known about the toxicity of individual components and the concentrations of the contaminants are higher, the UF might be set at 10. The greater the uncertainty involved (because of the lack of information) and the higher the concentrations of the contaminants, the higher the UF would be set.

Other Considerations

In keeping with the NRC report on complex mixtures (1988, p. 8), we have here defined "low" dose for carcinogens as one associated with a true relative risk of less than or equal to 1.01. Such a small relative risk is likely to be unmeasurable, and the "low" dose considered would be one estimated to produce a relative risk of 1.01 or less. The present subcommittee first considered tying the previous report's definition of "low" dose to detection limits, but decided not to do so, because this concept is related more to the physicochemical properties of an agent than to its biologic or toxic properties. Moreover, detection limits are likely to change as measurement technology improves. Further investigation into the magnitude of synergism as a function of dose might yield guidance here. Such investigation would involve the analysis of empirical data, alternative dose-response models, and additional laboratory experiments.

"Amount of information available" remains unavoidably ambiguous. Obviously, if perfect information were available about the existence and nature of any synergism, it could be included in a risk assessment. In the absence of complete knowledge, information about synergism among similar agents or about the mechanisms of toxicity of agents would be helpful.

In considering uncertainty factors, we recognize that a different type of uncertainty factor is already incorporated into the estimation of ADIs and ALs. Traditionally, this safety margin (usually a factor of 100) was provided to allow for possible interspecies and interindividual variations. Potential synergistic effects were not used in this development of safety margins. From a different perspective, however, the incorporation of a UF is not intended to trigger regulatory action every time contaminants are detected in drinking water sources. The hazard index in the new approach must be assessed in actual situations.

The above approaches apply to mixtures of systemic toxicants. For carcinogens, models that assume no threshold for response are recommended, and indeed they are used; hence, the constructs associated with ALs or ADIs do not apply. Mixtures of carcinogens are considered in Chapter 7.

For noncancer end points, dose-response models are not widely available (or accepted), because most toxicity evaluations have described effects in terms of single numerical values, such as the ED_{50} or NOAEL. That is reasonable when a single compound is under consideration, but it is not necessarily realistic for a combination. When it is known that the compounds interact to produce dose-additive effects, it might be useful to use the hazard index.

RESPONSE-SURFACE MODELING

Just as knowledge of the concentration-response curve is necessary to characterize the toxicity of a single agent, knowledge of the concentration-response surface is required to characterize the toxicity of a combination. Response-surface methods are mathematical and statistical techniques that have been developed to aid in the solution of particular types of problems in scientific and engineering processes (Box and Draper, 1987; Carter et al., 1983; Cornell, 1981; Khuri and Cornell, 1987; Myers, 1976). The report *ASA/EPA Conferences on the Interpretation of Environmental Data. 1. Current Assessment of the Combined Toxicant Effects* (EPA, 1986a) referred briefly to the potential applicability of response-surface methods for describing the effects of combinations of toxicants. The methods include experimental design, statistical inference, and mathematical techniques that, when combined, enable an experimenter to explore empirically the process of interest.

One important aspect of this approach is experimental design. It can be used to guide the generation of data suitable for estimating the parameters of a statistical model of the dose response relationship (NRC, 1988). Experimental designs have been developed to estimate the effects of each component of a mixture and the effects of interactions on the basis of a small number of experimental points. Such designs could be useful for indicating the presence or absence of low-order interactions, once an appropriate model has been determined. One design that permits the estimation of the interactions of interest and requires only a small number of experimental test groups is the 2^k-factorial experiment. Each of k factors is present at two levels, and each level of each factor is combined with each level of every other factor. Such a protocol requires 2^k observations. The data from such experiments allow the estimation of the effect of each of k toxicants and each of the interactions of 2, 3, ... , k toxicants (factors). If some interactions are of less interest, more compact experimental designs (fractional factorial designs), which require fewer groups of exposed subjects, are available (NRC, 1988).

The response-surface approach has been used to evaluate combinations of toxic substances (Kinne, 1972; Schnute and McKinnell, 1984; Voyer and Heltshe, 1984; Voyer et al., 1982). Its use seems appropriate in this setting and should be encouraged. The availability of computer graphics has permitted plotting of observed and fitted data as either the surface or contours of constant response. Such plots are often informative in assessing interactions among components. However, their use is limited to combinations of only a few toxicants. There is some expectation that graphic procedures can be developed to display the associated surfaces in higher dimensions; these efforts should be encouraged and supported.

CONCLUSIONS

Given the current availability of information about the presence and magnitude of synergism among contaminants in drinking water, no single risk-assessment approach can be justified scientifically for systemic toxicants. The possibility of synergism cannot be ignored, so approaches that consider only individual toxic agents are likely to be inadequate. Approaches based on additivity of response or additivity of doses that produce a given response are likely to be useful in many situations. In risk assessments of mixtures, these approaches should be limited to groups of agents that have similar mechanisms of action and act at the same biologic site.

Nevertheless, the possibility remains that some unexpected and important synergism (or antagonism) exists and that either additive model could seriously underestimate risk. Societal and policy concerns about the existence of such interactions could lead to the introduction of further uncertainty factors in risk assessment. Uncertainty factors should not be uniform, but should increase with increasing exposure and decrease with increasing knowledge about the agents in mixtures.

REFERENCES

ACGIH (American Conference of Governmental and Industrial Hygienists). 1983. TLVs: Threshold Limit Values for Chemical Substances and Physical Agents in the Work Environment, with Intended Changes for 1983–1984. Cincinnati, Ohio: American Conference of Governmental and Industrial Hygienists.

Box, G. E. P., and N. R. Draper. 1987. Empirical Model-Building and Response Surface. New York: John Wiley & Sons. 669 pp.

Carter, W. H., G. L. Wampler, and D. M. Stablein. 1983. Regression Analysis of Survival Data in Cancer Chemotherapy. New York: Marcel Dekker. 209 pp.

Cornell, J. A. 1981. Experiments with Mixtures: Designs, Models, and the Analysis of Mixture Data. New York: John Wiley & Sons. 305 pp.

EPA (U.S. Environmental Protection Agency). 1986a. ASA/EPA Conferences on the Interpretation of Environmental Data. 1. Current Assessment of the Combined Toxicant Effects, May 5–6, 1986.

EPA (U.S. Environmental Protection Agency). 1986b. Guidelines for the health risk assessment of chemical mixtures. Fed. Regist. 51(185):34014–34025.

Khuri, A. I., and J. A. Cornell. 1987. Response Surfaces: Designs and Analyses. New York: Marcel Dekker. 405 pp.

Kinne, O., ed. 1972. Environmental factors. Part 3 in Marine Ecology. A Comprehensive, Integrated Treatise on Life in Oceans and Coastal Waters, Vol. 1. New York: Wiley Interscience.

Murphy, S. D., R. L. Anderson, and K. P. du Bois. 1959. Potentiation of the toxicity of malathion by triorthotolyl phosphate. Proc. Soc. Exp. Biol. Med. 100:483–487.

Myers, R. H. 1976. Response Surface Methodology. 246 pp. [Distributed by R. H. Myers, Department of Statistics, Virginia Polytechnic Institute and State University, Blacksburg, Virginia.]

NRC (National Research Council). 1980. Principles of Toxicological Interaction Associated with Multichemical Exposures. Washington, D.C.: National Academy Press. 213 pp.

NRC (National Research Council). 1988. Complex Mixtures: Methods for In Vivo Toxicity Testing. Washington, D.C.: National Academy Press. 227 pp.

Schnute, J., and S. McKinnell. 1984. A biologically meaningful approach to response surface analysis. Can. J. Fish. Aquat. Sci. 41:936–953.

Smyth, H. F., C. S. Weil, J. S. West, and C. P. Carpenter. 1969. An exploration of joint toxic action: I. Twenty-seven industrial chemicals intubated in rats in all possible pairs. Toxicol. Appl. Pharmacol. 14:340–347.

Voyer, R. A., and J. F. Heltshe. 1984. Factor interactions and aquatic toxicity testing. Water Res. 18:441–447.

Voyer, R. A., J. A. Cardin, J. F. Heltshe, and G. L. Hoffman. 1982. Viability of embryos of the winter flounder *Pseudopleuronectes americanus* exposed to mixtures of cadmium and silver in combination with selected fixed sabinities. Aquat. Toxicol. 2:223–233.

Weiss, B. 1986. Emerging challenges to behavioral toxicology. Pp. 1–2 in Neurobehavioral Toxicology. Z. Annau, ed. Baltimore: Johns Hopkins University Press.

4

Assessment of Exposure to Organophosphorus Compounds, Carbamates, and Volatile Organic Chemicals

This chapter reviews methods for assessing exposure to organophosphorus compounds, carbamates, and volatile organic chemicals. It also provides orientation on the concentrations that could be encountered in water and on the resulting human exposures to them, and it indicates the importance of considering not only ingestion but other routes of exposure, such as skin contact and inhalation. Later chapters use these compounds to illustrate some toxicologic considerations of mixtures in drinking water.

Analytic methods for organophosphorus compounds and carbamates in water are well developed and standardized. An EPA method (EPA, 1984) for the determination of carbamates uses direct-injection high-performance liquid chromatography (HPLC). Less than 1 ml of a sample of filtered water is directly injected onto a reversed-phase HPLC column, and separation is achieved by gradient elution chromatography. The eluted compounds are hydrolyzed and then react with o-phthalaldehyde to form a fluorescent derivative, which is analyzed with a fluorescence detector. Although the detection limits for specific compounds in water vary, they are typically about 1 μg/liter for aldicarb, propoxur, carbaryl, carbofuran, and methomyl. Failure to detect a pesticide of the organophosphate or carbamate class known to be a potential contaminant of a specific water supply might not signify its absence; the pesticide might have hydrolyzed or undergone other chemical changes to other toxic substances, or the detection limit used in the analysis might have been too high.

An EPA method (EPA, 1986) for the assay of organophosphorus insecticides in water involves collection of a water sample, extraction of the sample with 1 liter of methylene chloride, concentration of the extract, and analysis

133

by gas chromatography with a capillary column and a nitrogen-phosphorus detector. For the typical organophosphorus insecticides diazinon, disulfoton, fonofos, and terbufos, the detection limit is about 1 mg/liter. The organophosphorus and carbamate insecticides hydrolyze at various rates in water, and in some cases their hydrolysis products are of toxicologic concern.

Maximum contaminant levels (MCLs) ranging from 0.002 to 0.75 mg/liter, depending on the compound, have been promulgated by EPA for the following volatile organic chemicals (VOCs) and became effective January 9, 1989: benzene, vinyl chloride, carbon tetrachloride, 1,2-dichloroethane, trichloroethylene, p-dichlorobenzene, 1,1-dichloroethylene, and 1,1,1-trichloroethane. There was already an MCL of 0.1 mg/liter for total trihalomethanes (THMs), a class of VOCs that includes chloroform, bromoform, and the mixed chlorobromomethanes. The THMs and VOCs are measured by purge-and-trap gas chromatography with estimated method detection limits (MDLs) of about 0.2–1.9 µg/liter (EPA, 1987a).

OCCURRENCE OF ORGANOPHOSPHORUS COMPOUNDS AND CARBAMATES IN DRINKING WATER

Several surveys have compiled data on concentrations of organophosphorus compounds and carbamates in surface waters, groundwaters, and, in some cases, completely treated ("finished") or well waters. Typically, their concentrations are 1 µg/liter or less (often undetectable), although substantially higher concentrations have been found in isolated instances. Kelley et al. (1986) showed that many commonly used pesticides, including carbamates and organophosphorus compounds, leach into groundwater. Typical concentrations in groundwater in Iowa were 0.5–2.0 µg/liter, although some wells had total pesticide concentrations of 20 µg/liter.

Data collected by various regulatory agencies are entered into the EPA surface-water and groundwater data base, including data on several pesticides. Such data have been summarized in health advisories prepared by EPA (1987b) for some of these pesticides. These and other data on various carbamate and organophosphorus insecticides are presented here to provide perspective on the concentrations of pesticides encountered in drinking water in several studies.

Aldicarb

In 17 of 106 wells sampled in California, aldicarb was detected at up to 14 parts per billion (ppb, equivalent to µg/l) (NRC, 1986). In the 15 states where aldicarb was found in groundwater, it was found typically at 1–50 µg/liter (Cohen at al., 1986).

Carbaryl

Carbaryl has been found in 61 of 522 surface water samples and 28 of 1,125 groundwater samples in eight states (EPA, 1987b). In samples with detectable concentrations, the 85th-percentile concentrations were 260 µg/liter in surface water and 10 µg/liter in groundwater. The highest concentrations were 180,000 µg/liter in surface water and 10 µg/liter in groundwater.

Carbofuran

Carbofuran has been found in groundwater in three states, typically at 1–50 µg/liter (Cohen et al., 1986).

Diazinon

Diazinon has been found in 13 wells (total number sampled not available) in California at up to 9 µg/liter (NRC, 1986, p. 20). It has been found in 7,230 of 23,227 surface-water samples and 115 of 3,339 groundwater samples in 46 states (EPA, 1987c). The 85th percentile of detectable concentrations was 0.2 mg/liter in surface water and 0.25 µg/liter in groundwater. The highest concentrations were 33,400 µg/liter in surface water and 84 µg/liter in groundwater.

Fonofos

In Iowa, fonofos was detected in about 2% of the samples at a typical concentration of 0.4 µg/liter; the maximum was 0.9 µg/liter (Kelley et al., 1986).

It was detected in groundwater in California at 0.01–0.03 µg/liter (EPA, 1987d).

Malathion

Malathion has been detected in five wells (total number sampled not available) in California at up to 23 µg/liter (NRC, 1986, p. 20).

Methyl parathion

Methyl parathion has been found in 1,402 of 29,002 surface-water samples and 25 of 2,878 groundwater samples in 22 states (EPA, 1987e). The 85th percentile of detectable concentrations was 1.2 µg/liter in surface water and

1 µg/liter in groundwater. The highest concentrations were 13 µg/liter in surface water and 1.6 µg/liter in groundwater.

Terbufos

In one study in Iowa, terbufos was found in 5% of the samples at a typical concentration of 5.4 µg/liter; the maximum was 12 µg/liter (Kelley et al., 1986). It has been found in 444 of 2,106 surface-water samples and 9 of 283 groundwater samples in five states (EPA, 1987f). The 85th percentile of detectable concentrations was 0.1 µg/liter in surface water and 3 µg/liter in groundwater. The highest concentrations were 2.3 µg/liter in surface water and 3 µg/liter in groundwater.

OCCURRENCE OF VOLATILE ORGANIC COMPOUNDS IN DRINKING WATER

There have been several surveys of THMs and VOCs in surface waters, groundwater, and finished water supplies. The straight lines fitted to results of an early EPA survey for THMs in finished water supplies of 80 U.S. cities are shown in Figure 4-1 (EPA, 1975). The median total concentration of THMs was about 20 µg/liter, and chloroform usually dominated the other THMs.

Results of several surveys of VOCs in surface waters and groundwater are summarized in Tables 4-1, 4-2, and 4-3. Although Table 4-1 shows very high concentrations of some specific organic chemicals (such as trichloroethylene at 35,000 µg/liter and 1,1,1–trichloroethane at more than 400,000 µg/liter), "more commonly, contamination is found at less than 10 µg/l with smaller percentages in the 10–100 µg/l and in the 100–1,000 µg/l range" (EPA, 1982). In one national survey, the VOCs most frequently found in finished groundwater supplies (other than the THMs) were trichloroethylene, 1,1,1-trichloroethane, tetrachloroethylene, *cis*- and *trans*-1,2-dichloroethylene, and 1,1-dichloroethane (Westrick et al., 1984). Table 4-2 summarizes the occurrences of the compounds detected at 186 randomly sampled sites serving more than 10,000 people each. Differences in median concentrations shown in Figure 4-1 and Table 4-2 arise largely out of the differences in sampled supplies. Figure 4-1 includes surface-water supplies, some of which are likely to be heavily chlorinated, whereas the data in Table 4-2 derive from groundwater sources, which are less likely to be chlorinated. The distribution of the summed concentrations of these VOCs, shown in Table 4-3, demonstrates that large systems were likely to exceed a summed concentration of 5.0 µg/liter slightly more frequently than small systems, as might be expected from purely statistical considerations. In both their random and nonrandom samplings, the median concentrations of specific compounds in

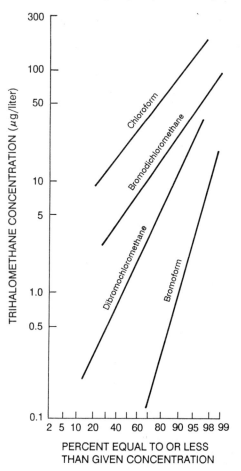

FIGURE 4-1 Frequency distribution of trihalomethane concentrations found in the National Organics Reconnaissance Survey (NORS) of halogenated organic compounds in drinking water in 80 U.S. cities. From EPA, 1975.

the positive samples ranged from about 0.2 to 9 µg/liter. One can conclude that VOCs other than THMs are normally found at concentrations of less than 10 µg/liter—and often less than 1 µg/liter—in finished groundwater supplies.

ROUTES OF HUMAN EXPOSURE TO CHEMICALS IN DRINKING WATER

The usual estimates of exposure to contaminants in drinking water are based on ingestion and are calculated from the standard of 2-liters/day ingestion of water by a 70-kg man. Ingestion of 2 liters/day is used to develop MCLs when the dose-response relationships are known. There has been some

TABLE 4-1   Occurrence of Volatile Organic Chemicals in Finished
Drinking Water[a]

Compound	Survey[b]	No. Samples	No. Positive	Concentration Range, μg/liter
Trichloroethylene	State data[c]	2,894	810	Trace–35,000
	NOMS[d]	113	28	0.2–49.0
	NSP[d]	142	36[e]	Trace–53.0
	CWSS[d]	452	15	0.5–210
Tetrachloroethylene	State data[c]	1,652	231	Trace–3,000
	NOMS[d]	113	48	0.2–3.1
	NSP[d]	142	24[e]	Trace–3.2
	CWSS[d]	452	22	0.5–30
Carbon tetrachloride	State data[c]	1,659	166	Trace–170
	NOMS[d]	113	14	0.2–29
	NSP[d]	142	37[e]	Trace–30
	CWSS[d]	452	9	0.5–2.8
1,1,1-Trichloroethane	State data[c]	1,611	370	Trace–401,300
	NOMS[d]	113	19	0.2–1.3
	NSP[d]	142	32[e]	Trace–21
	CWSS[d]	452	19	0.5–650
1,2-Dichloroethane	State data[c]	1,212	85	Trace–400
	NOMS[d]	113	2	0.1–1.8
	NSP[d]	142	2[e]	Trace–4.8
	CWSS[d]	451	4	0.5–1.8
Vinyl chloride	State data[c]	1,033	73	Trace–380
	NOMS[d]	113	2	0.1–0.18
	NSP[d]	142	7[e]	Trace–76
	CWSS[d]	[f]	[f]	[f]

[a]From EPA, 1982.

[b]NOMS, National Organics Monitoring Survey, 1976–1977; NSP, National Screening Program,
1977–1981; CWSS, Community Water Supply Survey, 1978 (EPA, 1987g).

[c]All groundwater sources; aggregated from various state reports on local contamination problems.

[d]Surface-water and groundwater sources.

[e]Tentative identification by single-column gas chromatography.

[f]Compound not surveyed.

effort to determine the effects of variability in the quantities of water ingested
(Gillies and Paulin, 1983). Results of studies in Canada, Great Britain, The
Netherlands, and New Zealand indicate that mean daily intakes, including
those of beverages made with tap water, range from 0.96 to 1.34 liters/day.
In one of the New Zealand studies, the mean intake of 960 ml/day had a
standard deviation of 570 ml/day. A more recent study in the United States
indicated higher water intake (Ershow and Cantor, 1986). Thus, in estimating
the toxic effects of ingestion of water that contains pesticides, one should
consider both the mean and the variability in intake in the population of
interest. Most attention should be paid to persons with the highest intake.

Attention has recently shifted toward exposure to chemical contaminants of tap water by routes other than ingestion, including skin contact with and inhalation of chemicals that volatilize indoors and can be inhaled at the point of water use or elsewhere or in moves from room to room. Very few data are available on cutaneous exposure, although Brown et al. (1984) estimated that bathing can cause exposures in the same range as those caused by the daily ingestion of 2 liters of the same water.

Field measurements, experimental studies, and models have measured the volatilization of chemicals found in water used for various purposes, including showering and bathing (Andelman, 1985; McKone, 1987). Estimates of the resulting inhalation exposures vary, but suggest that showering can cause about as much exposure as ingestion and that exposure due to all water uses can be substantially higher than that due to direct ingestion, in part because the intake of air (about 20,000 liters/day) is about 10^4 times that of water.

How quickly and how completely a pesticide or other contaminant in

TABLE 4-2   Summary of Occurrences of Volatile Organic Chemicals at 186 Randomly Sampled Groundwater Sites Serving More Than 10,000 Persons Each[a]

Compound	Quantitation Limit, µg/liter	Occurrences No.	Occurrences %	Median of Positives, µg/liter	Maximum, µg/liter
Vinyl chloride	1.0	1	0.5	1.1	1.1
1,1-Dichloroethylene	0.2	5	2.7	0.28	2.2
1,1-Dichloroethane	0.2	8	4.3	0.54	1.2
cis- and trans-1,2-Dichloroethylene	0.2	13	7.0	1.1	2.0
1,2-Dichloroethane	0.5	3	1.6	0.57	0.95
1,1,1-Trichloroethane	0.2	15	8.1	1.0	3.1
Carbon tetrachloride	0.2	10	5.4	0.32	2.8
1,2-Dichloropropane	0.2	5	2.7	0.96	21
Trichloroethylene	0.2	21	11.3	1.0	78
Tetrachloroethylene	0.2	21	11.3	0.52	5.9
Benzene	0.5	2	1.1	9.0	15
Toluene	0.5	2	1.1	2.6	2.9
Ethylbenzene	0.5	1	0.5	0.74	0.74
Bromobenzene	0.5	1	0.5	1.7	1.7
m-Xylene	0.2	2	1.1	0.46	0.61
o + p-Xylene	0.2	2	1.1	0.59	0.91
p-Dichlorobenzene	0.5	3	1.6	0.66	1.3
Chloroform	0.2	106	57.0	1.6	300
Bromodichloromethane	0.2	101	54.3	1.6	71
Dibromochloromethane	0.5	96	51.6	2.9	59
Dichloroiodomethane	1.0	3	1.6	1.8	4.1
Bromoform	1.0	57	30.6	3.8	50

[a]Adapted from Westrick et al., 1984, with permission.

TABLE 4-3   VOC Concentrations in Random Samples of Finished Groundwater[a]

Summed Concentrations of VOCs, μg/liter	Water Supplies with Summed Concentrations of VOCs Exceeding Value Shown at Left			
	Systems Serving Up to 10,000 Persons		Systems Serving More Than 10,000 Persons	
	No.	%	No.	%
Quantitation limit[b]	47	16.8	52	28.0
1.0	20	7.1	26	14.0
5.0	8	2.9	12	6.5
10	5	1.8	7	3.8
50	1	0.4	1	0.5
100	0	0	0	0

[a]Adapted from Westrick et al., 1984, with permission.

[b]Quantitation limits not same for all compounds. In most cases, quantitation limit is either 0.2 μg/l or 0.5 μg/l. This difference in quantitation limits can confuse interpretation of data somewhat, so results of survey should be viewed with differing quantitation limits in mind. "Occurrence" is any specific finding at, or in excess of, the quantitation limit.

drinking water will volatilize depends on its physical and chemical properties, including its solubility in water, its vapor pressure, its Henry's law constant (H), and its coefficient of diffusion in water at the water-air interface (Andelman, 1985), as well as physical characteristics of the water, such as temperature, agitation, and spraying. The constant H is equal to the ratio of the equilibrium concentration in air to the concentration in aqueous solution.

The vapor pressures of carbamates and organophosphorus compounds are low, as are their H values. Other aqueous insecticides, such as dieldrin and aldrin, can readily volatilize from water-air interfaces, although probably at lower rates than compounds (such as benzene and toluene) that have higher H values (Mackay and Leinonen, 1975). Polychlorinated biphenyls (such as Arochlor 1242) and chlordane volatilized from water surfaces at about 20%–30% of the rate of oxygen in reaeration studies, but dieldrin, which has a substantially lower H value, volatilized at only 1%–5% of the oxygen reaeration rate (Atlas et al., 1982). The H values of specific organophosphorus and carbamate compounds vary, but many are low. On the basis of their water solubilities and vapor pressures, one can calculate H values at room temperature of 2×10^{-6} and 8×10^{-9} atm · m^3/mol for aldicarb and carbofuran, respectively. Comparable values for dieldrin, aldrin, and Arochlor 1242 at room temperature are 2×10^{-7}, 1×10^{-5}, and 6×10^{-4} atm · m^3/mol (Mackay and Leinonen, 1975).

In contrast to the organophosphorus and carbamate compounds, the typical VOC or THM has a relatively large H value at room temperature, generally in the range of 10^{-2}–10^{-3} atm · m³/mol (Roberts and Dandliker, 1983). The mass-transfer rate constants for the VOCs and THMs are also substantially higher than those for the organophosphorus and carbamate compounds, typically about 60% of that for oxygen reaeration. Thus, one would expect that substantial fractions of these components would volatilize during typical indoor water uses and thereby contribute to inhalation exposures, especially of the person at the point of use, but perhaps of others if the volatile constituents are disseminated by air movement. As has been shown for trichloroethylene and chloroform, volatilization from both showers and baths is substantial, usually greater than 50% and sometimes as high as 90%, depending on temperature, air flow, and the geometry of the water system (Andelman, 1985; Andelman et al., 1986, 1987). For the VOCs and THMs with higher H values, Henry's law equilibrium is generally not attained, so mass transfer at the water-air interface often limits the rate and extent of volatilization. A recent attempt to determine whether a surrogate chemical, sulfur hexafluoride, could be used to estimate the volatilization of such constituents associated with indoor water uses (Giardino et al., 1988) was encouraging, but additional research is required.

CONSIDERATIONS OF TOTAL EXPOSURE

The principal focus of this report is the assessment of toxicity associated with exposure to mixtures of chemicals in drinking water. But other media, such as food, are also potential sources of exposure. Exposures related to nonwater sources, such as exposure to polycyclic aromatic hydrocarbons in ambient air, can be much greater than those related to water. When human toxicity associated with exposure through water is assessed, combined exposures through other media have the potential for raising an apparently low exposure through water to the point where a toxic threshold is exceeded or, in the case of a carcinogen, a risk is increased.

An early multimedia-exposure analysis for some of the chemicals considered here addressed the multiple routes and variability of uptake of chloroform and carbon tetrachloride (NRC, 1978). The exposure data on air, water, and food in that analysis were often meager and not precise, but the analysis did use what was known about variability in absorption after ingestion or inhalation. Table 4-4 shows three hypothetical scenarios for uptake (based on exposure and absorption) of carbon tetrachloride and chloroform from water, food, the atmosphere, and the three together. It appears that most exposure to chloroform at typical levels is by water, whereas air is typically more important for carbon tetrachloride. However, in any given instance, almost any route can dominate, so it is essential to consider all sources when one is assessing individual exposure to a specific chemical and the associated risk.

TABLE 4-4   Adult Human Male Uptake (Based on Exposure and Intake) of Carbon Tetrachloride (CCl$_4$) and Chloroform (CHCl$_3$) from Environmental Sources[a]

	Uptake, mg/year	
Source	CCl$_4$	CHCl$_3$
At minimum exposure levels[b]		
Water and water-based drinks	0.73	0.037
Atmosphere	3.60	0.41
Food	0.21	0.21
Total	4.54	0.66
At typical exposure levels[c]		
Water and water-based drinks	1.78	14.90
Atmosphere	4.80	5.20
Food	1.12	2.17
Total	7.70	22.3
At maximum exposure levels[d]		
Water and water-based drinks	4.05	494
Atmosphere	618	474
Food	7.33	16.4
Total	629	984

[a]Adapted from NRC, 1978, pp. 180–181.

[b]Minimum exposure and minimum intake for all sources.

[c]Typical conditions assumed. For CCl$_4$: water and water-based drinks, exposure at 0.0025 mg/liter and reference-man intake; atmosphere, average of typical minimum and maximum absorption; food, average exposure and intake. For CHCl$_3$: water and water-based drinks, median exposure and reference-man intake; atmosphere, average of typical minimum and maximum absorption; food, average exposure and intake.

[d]Maximum exposure and maximum intake for all sources.

CONCLUSIONS AND RECOMMENDATIONS

If joint exposure to THMs or to all VOCs with roughly equivalent potency could be considered to have additive toxic effects, it would be useful to have an analytic method for monitoring purposes that could be used as a measure of the total concentration of members of the group. For example, the sum of the volatile organohalide concentrations could be measured with a single instrument, even though it would measure a group of compounds—such as vinyl chloride, carbon tetrachloride, 1,2-dichloroethane, trichloroethylene, p-dichlorobenzene, 1,1-dichloroethylene, and 1,1,1-trichloroethane—with widely varied toxic effects. The potential deficiency of such a method is that other, possibly harmless volatile organohalide compounds in the water sample would also be detected. In the case of the organophosphorus compounds, it is unlikely that a single simple analytic measurement with a gas chromatographic phosphorus detector can be developed that would depend only on the presence of organophosphorus insecticides, because other, unidentified phosphorus compounds could be present. The HPLC method described earlier might be more successful in this regard, but it would have to be shown that the derivatization procedure is specific to the hydrolysis products of the carbamates of interest and that other naturally occurring chemicals and their hydrolysis products would not introduce serious inaccuracies.

If a simple analytic process could be developed to provide a summary measure of the concentrations of an entire class of toxicologically similar constituents in drinking water, it is likely that it would also detect other, potentially confounding constituents in the water.

In assessing the toxic impacts of individual or mixed constituents of drinking water, it is essential to consider all forms of exposure to the constituents, including exposure through water, soil, air, and food. For contaminated drinking water, exposure by inhalation, skin contact, and ingestion should be assessed. Whether organophosphorus and carbamate compounds volatilize to a substantial extent during domestic indoor water uses should be determined, so that total exposures to these chemicals can be assessed. For the THMs and VOCs, exposure due to volatilization can be substantial and should be considered in assessing human toxic impact and risk.

REFERENCES

Andelman, J. B. 1985. Inhalation exposure in the home to volatile organic contaminants of drinking water. Sci. Total Environ. 47:443–460.

Andelman, J. B., S. M. Meyers, and L. C. Wilder. 1986. Volatilization of organic chemicals from indoor water uses. Pp. 323–330 in Chemicals in the Environment, J. N. Lester, R. Perry, and R. M. Sterritt, eds. London: Selper.

Andelman, J. B., L. C. Wilder, and S. M. Meyers. 1987. Indoor air pollution from volatile chemicals in water. Pp. 37–41 in Proceedings of the 4th International Conference on Indoor Air Quality and Climate, Vol. 1. West Berlin, August 1987.

Atlas, E., R. Foster, and C. S. Giam. 1982. Air-sea exchange of high molecular weight organic pollutants: Laboratory studies. Environ. Sci. Technol. 16:283–286.

Brown, H. S., D. R. Bishop, and C. A. Rowan. 1984. The role of skin absorption as a route of exposure for volatile organic compounds (VOCs) in drinking water. Am. J. Public Health 74:479–484.

Cohen, S. Z., C. Eiden, and M. N. Lorber. 1986. Monitoring ground water for pesticides. Pp. 170–196 in Evaluation of Pesticides in Ground Water. ACS Symposium Series Vol. 315. Washington, D.C.: American Chemical Society.

EPA (U.S. Environmental Protection Agency). 1975. Preliminary Assessment of Suspected Carcinogens in Drinking Water, and Appendices. A Report to Congress. Washington, D.C.: U.S. Environmental Protection Agency.

EPA (U.S. Environmental Protection Agency). 1982. National revised primary drinking water regulations, volatile synthetic organic chemicals in drinking water; advanced notice of proposed rulemaking (March 4, 1982). Fed. Regist. 47(43):9350–9358.

EPA (U.S. Environmental Protection Agency). 1984. Method 531. Measurement of *N*-Methyl Carboxylamines and *N*-Methylcarbamates in Drinking Water by Direct Aqueous Injection HPLC with Post Column Derivatization. Cincinnati, Ohio: Environmental Monitoring and Support Laboratory, U.S. Environmental Protection Agency.

EPA (U.S. Environmental Protection Agency). 1986. Method 1. Determination of Nitrogen- and Phosphorus-Containing Pesticides in Ground Water by GC/NPD. January 1986 Draft Report. Cincinnati, Ohio: Environmental Monitoring and Support Laboratory, U.S. Environmental Protection Agency.

EPA (U.S. Environmental Protection Agency). 1987a. Volatile Organic Chemicals: Methods and Monitoring Document. Cincinnati, Ohio: Technical Support Division, U.S. Environmental Protection Agency.

EPA (U.S. Environmental Protection Agency). 1987b. Carbaryl, August, 1987. Health Advisory. Draft. Washington, D.C.: Office of Drinking Water, U.S. Environmental Protection Agency. 20 pp.

EPA (U.S. Environmental Protection Agency). 1987c. Diazinon, August, 1987. Health Advisory. Draft. Washington, D.C.: Office of Drinking Water, U.S. Environmental Protection Agency. 19 pp.

EPA (U.S. Environmental Protection Agency). 1987d. Fonofos, August 1987. Health Advisory. Draft. Washington, D.C.: Office of Drinking Water, U.S. Environmental Protection Agency. 16 pp.

EPA (U.S. Environmental Protection Agency). 1987e. Methyl Parathion, August 1987. Health Advisory. Draft. Washington, D.C.: Office of Drinking Water, U.S. Environmental Protection Agency. 24 pp.

EPA (U.S. Environmental Protection Agency). 1987f. Terbufos, August, 1987. Health Advisory. Draft. Washington, D.C.: Office of Drinking Water, U.S. Environmental Protection Agency. 15 pp.

EPA (U.S. Environmental Protection Agency). 1987g. Drinking water; substitution of contaminants and priority list of additional substances which may require regulation under the Safe Drinking Water Act. Fed. Regist. 52(130):25720–25734.

Ershow, A., and K. P. Cantor. 1986. Population-based estimates of water intake. Fed. Proc. 45:706.

Giardino, N., J. B. Andelman, J. E. Borrazzo, and C. I. Davidson. 1988. Sulfurhexafluoride as a surrogate for volatilization of organics from indoor water uses. J. Air Pollut. Control Assoc. 3:278–280.

Gillies, M. E., and H. V. Paulin. 1983. Variability of mineral intakes from drinking water: A possible explanation for the controversy over the relationship of water quality to cardiovascular diseases. Int. J. Epidemiol. 12:45–50.

Kelley, R., G. R. Hallberg, L. G. Johnson, R. D. Libra, C. A. Thompson, R. G. Splinter, and M. G. DeTroy. 1986. Pesticides in ground water in Iowa. J. Natl. Well Water Assoc. August:622–647.

Mackay, D., and P. J. Leinonen. 1975. Rate of evaporation of low-solubility contaminants from water bodies to atmosphere. Environ. Sci. Technol. 9:1178–1180.

McKone, T. E. 1987. Human exposure to volatile organic compounds in household tap water: The indoor inhalation pathway. Environ. Sci. Technol. 21:1194–1201.

NRC (National Research Council). 1978. Chloroform, Carbon Tetrachloride and Other Halomethanes: An Environmental Assessment. Washington, D.C.: National Academy of Sciences. 294 pp.

NRC (National Research Council). 1986. Pesticides and Groundwater Quality: Issues and

Problems in Four States. Written by Patrick W. Holden. Washington, D.C.: National Academy Press. 124 pp.

Roberts, P. V., and P. Dandliker. 1983. Mass transfer of volatile organic contaminants from aqueous solution to the atmosphere during surface aeration. Environ. Sci. Technol. 17:484–489.

Westrick, J. J., J. W. Mello, and R. F. Thomas. 1984. The groundwater supply survey. J. Am. Water Works Assoc. 76(5):52–59.

5

Acetylcholinesterase Inhibitors: Case Study of Mixtures of Contaminants with Similar Biologic Effects

Acetylcholine, a neurotransmitter normally present in many parts of the nervous system, is hydrolyzed by the enzyme acetylcholinesterase. Chemicals that inhibit the action of acetylcholinesterase at doses or concentrations substantially lower than those required for other kinds of biologic effects are pharmacologically classified as anticholinesterases. Anticholinesterases appear to mimic the stimulation of cholinergic nerves or receptors in the central and peripheral nervous systems.

This chapter discusses the toxicity and interactions of the two groups of chemicals most often associated with anticholinesterase activity—the organic triesters of phosphoric (P=O) or phosphorothioic (P=S) acid (i.e., organophosphorus compounds) and several carbamates (esters of carbamic acid). Chemicals in both groups are widely used as insecticides. However, not all organophosphorus triesters or all carbamates are insecticidal, nor can all of them be classified as anticholinesterases. It is important to keep that in mind to avoid overgeneralizing when discussing these chemicals (either as pharmacologic classes or as chemical classes) with respect to their toxic actions and the regulatory decisions concerning them.

Although organophosphorus compounds and carbamates are often used as insecticides, the anticholinesterases have other applications as well. The drug physostigmine, obtained from the calabar bean, is an aromatic carbamate ester that was first used therapeutically in 1877 in the treatment of glaucoma and still has some use for this purpose. Other related carbamates (such as neostigmine and edrophonium) and a few organophosphorus esters (such as diisopropyl phosphorofluoridate, octamethyl pyrophosphoramide, and echothiophate) have been used clinically to stimulate the smooth muscles of the ileum or the urinary bladder in paralytic ileus and atony of the urinary bladder, to decrease intraocular tension in glaucoma, and to overcome the muscular weakness and rapid fatigability of skeletal muscle in myasthenia gravis. Except for physostigmine and neostigmine, however, the clinical uses of anticholinesterases are very limited.

Large stocks of organic triesters of phosphoric acid that are potent anticholinesterases have been stored for potential use as chemical warfare agents. In fact, the use of the organophosphorus compounds in agriculture, as well as clinical medicine, was an outgrowth of the chemical-warfare research during World War II. The agents could become environmental contaminants in areas where they have been tested in military field operations or as a result of leakage from disposal sites.

In summary, the anticholinesterases are primarily in two chemical classes: the organic triesters of phosphoric or phosphorothioic acid and the carbamates. The chemicals have been developed for use as chemical-warfare agents, as insecticides, and in clinical medicine; their most probable source as surface-water and groundwater contaminants is insecticides.

TOXICITY

The known toxic effects of the anticholinesterases are predominantly the acute effects elicited by single doses (Murphy, 1986). Both the organophosphorus and carbamate classes of anticholinesterases contain compounds whose acute lethal dosages range from a few milligrams per kilogram to greater than a gram per kilogram (Murphy, 1986). The manifestation of cumulative toxic action is generally the same as that of the action produced by a large single dose. The effects usually appear in several organs, because acetylcholine accumulates at the synapses of cholinergic nerves when acetylcholinesterase is inhibited and has muscarinic, nicotinic, and central nervous system actions. Some organophosphorus compounds or carbamates have other toxic actions, such as carcinogenicity or teratogenicity, that are not associated with the anticholinesterase action. The chronic effects of the compounds are generally compound-specific and cannot be defined as characteristic of the class. A possible exception is delayed peripheral neuropathy, known as organophosphorus-compound-induced delayed neurotoxicity (OPIDN), which reflects a primary axonal degeneration caused by some of the organophosphorus triesters. Many, but not all, organophosphorus triesters that produce OPIDN are also strong inhibitors of acetylcholinesterase. The inhibition of acetylcholinesterase appears to be unrelated to the mechanism of production of OPIDN. In fact, in many cases the capacity of a chemical to produce delayed OPIDN has been discovered only when doses greater than the dose that would be lethal owing to anticholinesterase or cholinergic action could be administered to test animals (usually fully grown hens). Such testing is possible with the use of atropine, which protects the muscarinic receptors from accumulating acetylcholine.

The signs and symptoms of acute poisoning by the anticholinesterases usually reflect the actions of acetylcholine at muscarinic receptors in smooth muscle, the heart, and exocrine glands. They include tightness in the chest, wheezing, and increases in bronchial secretion, salivation, lacrimation, sweating, and gastrointestinal tone and peristalsis, with the consequent development of nausea, vomiting, abdominal cramps, and diarrhea. There can be slowing of the heart (which can progress to heart block), infrequent and involuntary urination, and constriction of the pupils.

The signs of poisoning by anticholinesterases that are associated with stimulation of nicotinic receptors include contractions of skeletal muscle, leading first to scattered and then generalized fasciculations and finally to muscular weakness and ultimately paralysis. The skeletal muscles include the muscles of respiration, and their paralysis is often the immediate cause of death. Nicotinic actions also include those at autonomic ganglia; in severe intoxication, the effects at synapses in the autonomic ganglia can mask the more usual muscarinic effects.

Accumulation of acetylcholine in the central nervous system can be responsible for the tension, anxiety, restlessness, insomnia, headaches, emotional instability and neurosis, excessive dreaming and nightmares, apathy, confusion, and forgetfulness reported by persons poisoned with anticholinesterases. Generally, if a person survives an episode of acute poisoning, recovery is complete. However, chronic sequelae involving the central nervous system (such as forgetfulness, dreaming, and electroencephalographic changes) have been reported to persist for a long time.

All the signs and symptoms described above can result from a single dose of an anticholinesterase agent that passes the blood-brain barrier, gains access to cells in the central nervous system, and acts at synapses of peripheral nerves. Smaller doses can be tolerated without these signs, but frequent repetition of the smaller doses can lead eventually to their onset when the accumulated inhibition of acetylcholinesterase allows acetylcholine to reach an excessive concentration. Results of most studies in experimental laboratory animals, as well as clinical observations and research, have indicated that inhibition of acetylcholinesterase must be substantial (e.g., 50% before signs typical of acute poisoning become manifest. However, there is some reason to suspect that subtle and unrecognized effects in the central nervous system can occur with smaller degrees of inhibition (Roney et al., 1986).

Mechanisms

The actual manifestation of acetylcholinesterase-related poisoning is mediated by the accumulation of the endogenous neurotransmitter acetylcholine, which affects receptors in effector organs and the brain.

The organophosphorus triesters phosphorylate the chemically active sites of the enzyme acetylcholinesterase, and the carbamate insecticides carbamylate the same sites. In a sense, they both act as alternative substrates, with the normal substrate being acetylcholine. The acetylated site dissociates very rapidly, the carbamylated site dissociates slowly, and the phosphorylated site dissociates even more slowly. Hence, cholinesterase inhibition by the phosphate compounds generally lasts longer than that by the carbamate compounds. With the carbamate compounds, spontaneous reversal of cholinesterase inhibition occurs when the excessive inhibitor has been metabolized or otherwise removed—generally within a few minutes to a few hours. The phosphorylated acetylcholinesterase of the organophosphorus compounds tends to be much more stable, and spontaneous dephosphorylation and regeneration of the uninhibited enzyme can take many hours to several days. On exposure to organophosphorus compounds, a portion of the inhibited enzyme is never spontaneously dephosphorylated; hence, some inhibition of cholinesterase can last for weeks, or until synthesis of new enzyme fully restores normal activity. This implies that the recovery mechanism is more complex than simple first-order kinetics.

Knowledge of the primary biochemical lesion associated with poisoning by the two classes of compounds has resulted in a convenient means for following the course of poisoning—measurement of the cholinesterase activity in erythrocytes or plasma. Inhibition of acetylcholinesterase activity in erythrocytes is thought to reflect the course of inhibition and reversal of inhibition in nerve tissue. A problem in the assay of carbamate compounds with cholinesterase inhibition is that rapid spontaneous reversal of inhibition can occur in vitro after blood has been drawn. It can also occur in vivo. Thus, if a cholinesterase assay of blood from a severely poisoned person is not conducted very promptly, it might fail to confirm carbamate poisoning. The slower reversibility of the inhibition caused by the organophosphorus compounds lessens this diagnostic problem.

There is substantial evidence from studies in laboratory animals, as well as some indication from studies in humans, that repeated exposures to subacute doses of organophosphorus compounds or persistent exposures to carbamate anticholinesterases can cause a form of tolerance to these compounds or a refractoriness of direct-acting cholinergic agonists. The phenomenon appears to be due to a reduction in the responsiveness of the cholinergic receptor system—a reduction in the density of cholinergic receptors—and it has been demonstrated for both muscarinic and nicotinic cholinergic receptors (Bombinski and DuBois, 1958; Brodeur and DuBois, 1964; Costa et al., 1982a,b; Schwab et al., 1981). Induction of such tolerance appears to require a prolonged inhibition of acetylcholinesterase (which results in a prolonged increase in acetylcholine at the receptor site) and thus has generally been reported for only the more persistent anticholinesterase compounds. If mixtures of anticholinesterases act to prolong inhibition, it is conceivable that early additivity or even synergism might give way to tolerance or apparent antagonism with prolonged exposures. If the manifestation of cholinergic effects were used as the end point, the tolerance might be interpreted as apparent antagonism; but if acetylcholinesterase inhibition were the end point, antagonism would not be apparent.

The preponderance of reported evidence indicates that all whole-organism signs and symptoms caused by anticholinesterases are preceded or accompanied by a significant inhibition of acetylcholinesterase. However, there are reports that some behavioral changes have persisted long after cholinesterase activity has returned to normal. Furthermore, some (sparse) experimental data (Roney et al., 1986) indicate that tests of subtle learned behaviors in laboratory animals can be altered with very little reduction in blood cholinesterase.

Some organophosphorus and carbamate compounds are not strong inhibitors of acetylcholinesterases and are not properly classed with the anticholinesterases. Those chemicals have their own cholinesterase-independent toxic action and cannot be grouped with the anticholinesterases for regulatory purposes. Examples are the triaryl phosphates, including the classic peripheral neurotoxic compound tri-*o*-cresylphosphate, the dithiocarbamate fungicides, and some of the carbamate herbicides.

Metabolism-Toxicity Relationships

ORGANOPHOSPHORUS COMPOUNDS

The broad class of organophosphorus anticholinesterases includes several types of compounds. Some are esters of phosphoric acid, $(RO)_3—P{=}O$, some are esters of phosphorothioic acid, $(RO)_3—P{=}S$, and a few are phosphonates and phosphoroamidates. The phosphoric acid triesters ($P{=}O$ compounds) are active insecticides and are generally direct inhibitors of acetylcholinesterase. That is, when added to a solution of purified or partially purified acetylcholinesterase or to a crude homogenate or extract of tissue containing acetylcholinesterase, they exhibit potent in vitro anticholinesterase action, often at concentrations of $10^{-10}–10^{-7}\ M$. When the derivatives of phosphorothioic acid ($P{=}S$ compounds) are added to a cholinesterase-containing preparation, they are not strong direct inhibitors of cholinesterase. They are, however, active in vivo at relatively low doses, and tissues taken from animals poisoned with low doses of these compounds have greatly reduced concentrations of acetylcholinesterase.

It is now well established that the $P{=}S$ compounds must be activated by other enzymes in the body to a form that is highly reactive with the acetylcholinesteraseactive center. The most common case with the $P{=}S$ derivatives is conversion to the $P{=}O$ (phosphate) form of the triester, which is a direct

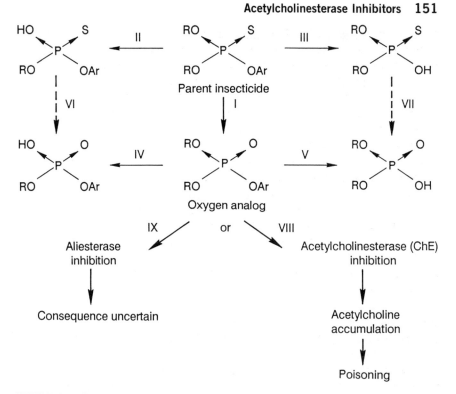

FIGURE 5-1   General scheme of metabolism and mechanism of toxic action of dialkyl aryl phosphorothioates. From Murphy, 1980, with permission.

inhibitor of acetylcholinesterase (I in Figure 5-1). Hence, factors that alter the rate of metabolism of these indirect inhibitors to their directly inhibiting forms can alter the toxicity of the compounds.

Furthermore, many of the P=O compounds can be attacked directly by hydrolases, sometimes called A-esterases, that split the $(RO)_2P$—O—OAr bond (V in Figure 5-1) and result in products that do not inhibit acetylcholinesterase. The P=S compounds are generally resistant to those hydrolases and are hydrolyzed only after they are converted to their oxygen analogue (I and V in Figure 5–1). It is now well established that most of the P=S compounds can be oxidatively cleaved to a phosphorothioic diester and an aryl hydroxy group (III in Figure 5-1). The oxidative cleavage step and the hydrolytic step (V) are detoxifying, and the products of these pathways do not inhibit acetylcholinesterase. To complicate the matter, the enzyme that oxidatively cleaves (III in Figure 5-1) and oxidatively desulfurates (I in Figure 5-1) phosphorothioates is either the same enzyme or two very closely related enzymes. They are classed as mixed-function oxidases, which are well known to be induced and inhibited by many other compounds (Murphy, 1986). In addition to the oxidative activation and the oxidative and hydrolytic inactivation of the phosphorothioate compounds, some of the compounds are also detoxified via glutathione alkyltransferases that remove an alkyl group from the phosphate and render it inactive as an anticholinesterase (II and IV in Figure 5-1). In a few cases, organophosphorus compounds have also been demonstrated to be dealkylated by oxidative enzymes. That is also a detoxifying step and is probably another form of a mixed-function oxidase.

In summary, oxidative metabolic pathways can either activate or detoxify the phosphorothioate insecticides. Such factors as inducers or inhibitors of mixed-function oxidases and competition by other compounds for the reactive sites in the mixed-function oxidases can alter the quantity of the active direct inhibitor of acetylcholinesterase at critical sites in nerve tissue (VIII in Figure 5-1) that will be present with any given dose at any given time. In addition, a different set of enzymes, the soluble glutathione alkyltransferases, might also detoxify some of the compounds. One further means of detoxification is the reaction of the organophosphorus compounds with other noncritical enzymes (IX in Figure 5-1) that can serve as a sink to divert the active phosphates from critical sites and spare acetylcholinesterase.

The complex multiple pathway of metabolism renders it extremely difficult to predict the possibility or quality of toxic interactions among mixtures of the organophosphorus compounds. In addition, conditions that might predict results with one compound or homologues of a compound might not apply for other, equally closely related compounds. For example, it has been demonstrated that inhibition of mixed-function oxidases by piperonyl butoxide or SKF 525A in mice moderately increases the toxicity of ethyl parathion, but protects strongly against the toxicity of methyl parathion (Levine and Murphy, 1977a,b). The reason appears to be that the alternate pathway for detoxification through glutathione transferase is effective for methyl parathion, but not for ethyl parathion.

A few of the organophosphorus compounds have chemical groups that can be attacked by other enzymes. Notable among them are chemicals that have carboxyl ester or carboxy amide linkages. The ester linkages can be attacked by widely distributed carboxyl esterase or carboxy amidase in tissues. The action of those hydrolases generally leads to a loss of anticholinesterase action by the phosphate or carbamate that contains the groups. Hence, compounds (including many insecticidal and noninsecticidal organophosphorus compounds) that inhibit carboxyl esterases can increase the toxicity of other organophosphorus compounds whose anticholinesterase activity depends on an intact carboxyl ester or carboxy amide linkage (DuBois, 1969; Murphy, 1969). The best-known synergism of this kind is with malathion (Casida et al., 1963; Cohen and Murphy, 1971; Frawley et al., 1957; Murphy et al., 1959), which is usually considered a relatively safe insecticide. Many laboratory studies and a field accident (Baker et al., 1978) have demonstrated that malathion becomes much more toxic under conditions in which carboxyl esterases are inhibited.

Finally, it has been demonstrated that some organophosphorus triesters that are not always potent inhibitors of acetylcholinesterase can compete with other anticholinesterases for a noncritical group of enzymes, sometimes referred to as aliesterases (IX in Figure 5-1); these include nonspecific carboxyl esterases. The competition can block a sink of noncritical binding sites that normally act to spare acetylcholinesterase (the critical binding site) from being inhibited by the organophosphates. Synergism among some organophosphorus compounds might depend on such action (Fleisher et al., 1963; Lauwerys and Murphy, 1969; Murphy et al., 1976; Polak and Cohen, 1969).

CARBAMATES

The carbamate insecticides also have multiple pathways of metabolism, which are also predominantly oxidative and hydrolytic. Hydrolysis of the carbamate ester invariably reduces its anticholinesterase activity, but oxidative reactions that occur on the ring or alkyl portions of the carbamate insecticides can increase or decrease anticholinesterase activity. For example, in the case of the carbamate insecticide propoxur, hydrolysis of the carbamate ester linkage reduces the anticholinesterase potency by a factor of 100. With the carbamate ester intact, oxidative removal of the isopropoxy group reduces toxicity by a factor of only about 5–6, hydroxylation of the N-methyl group reduces toxicity by a factor of only 4, and hydroxylation of the aromatic ring without other changes actually increases the anticholinesterase potency by a factor of 3 (Oonnithan and Casida, 1968). There is much less information on the interactions that can occur between carbamates or between carbamates and phosphate anticholinesterases than there is on interactions between the phosphate anticholinesterases. However, a recent report (Takahashi et al., 1987) indicated that the toxicity of N-methyl carbamate compounds in mice can be increased by organophosphorus insecticides. The extent of synergism varied widely, from a factor of 2 to a factor of 15, depending on the organophosphorus compounds tested. The precise mechanism of the synergism is not entirely clear from the results of the study.

INTERACTIONS

Reported Data

The conceptual model that is usually applied to pesticides is the dose-additive model, and in the remainder of this chapter when the words synergism, antagonism, and interaction are used, they imply departures from dose additivity.

Over 30 years ago, Frawley and coworkers (1957) reported a marked synergism of two organophosphorus insecticides, ethyl p-nitrophenyl thionobenzenephosphonate (EPN) and malathion. For several years thereafter, the Food and Drug Administration required that all safety evaluations on all anticholinesterase insecticides for which food-residue tolerances were established include tests of the toxicity of combinations. Most of the tests conducted in response to the regulation were acute-toxicity tests that used simultaneous administration of two chemicals (binary mixtures).

DuBois (1961) reported on studies in which various combinations of 13 organophosphorus insecticides were tested for acute toxicity in rats. Toxicities of 21 pairs showed dose additivity, of 18 pairs were less than additive, and of 4 pairs were synergistic. Administration of half the LD_{50} of each of the two compounds in each pair, which should have led to 50% mortality, was followed by 100% mortality in 4 pairs. Three of the 4 pairs included malathion. A few more pairs involving newer compounds have since been shown to be synergistic in acute-toxicity tests. However, combinations of several organophosphorus insecticides that were incorporated into experimental diets at residue-tolerance limits did not show greater than additive toxicity in chronic feeding studies. The regulation requiring tests of anticholinesterase insecticides for synergism was lifted a few years after it was instituted, when investigators failed to demonstrate synergism at the residue-tolerance limits.

The likely mechanisms of synergism among organophosphorus insecticides have been reviewed by DuBois (1969) and Murphy (1969). Inhibition of detoxification by tissue carboxyl esterases and amidases and competition for nonvital binding sites that normally act as a buffer system to spare the vital acetylcholinesterase appeared to be the two major mechanisms involved in the synergism among organophosphorus compounds. One of the insecticides most often observed to be synergistic with other organophosphorus insecticides was malathion. Malathion, normally a relatively safe compound, is detoxified by carboxyl esterases that are inhibited by other organophosphorus insecticides (Murphy et al., 1959).

Clear evidence of the synergistic action of organophosphorus compounds in humans did not emerge until an incident among spraymen in a mosquito control program in Pakistan in the late 1970s resulted in several deaths and thousands of clinical poisonings. The incident was attributed to an increase in the toxicity of malathion due to interaction with other organophosphorus compounds that were strong inhibitors of carboxyl esterases and that constituted impurities in the malathion (Baker et al., 1978).

The other proposed mechanism of synergism is competition for noncritical binding sites. It has been suggested that the carboxyl esterases might represent one type of noncritical binding site. Compounds that are more potent inhibitors of carboxyl esterases than of cholinesterase might be the most likely to interact synergistically. The anticholinesterase action appears to be increased by a factor of 3–4 (i.e., the dose for an equitoxic effect is reduced to one-third or one-fourth) in acute doses of combinations of this type (DuBois, 1961; McCollister et al., 1959; Murphy, 1969, 1976).

Although the possibility of interactions among anticholinesterase compounds has been less studied, Takahashi et al. (1987) recently demonstrated increases in the toxicity of five N-methyl carbamates by simultaneous treatments or pretreatments with one-twentieth of the LD_{50} of some organophosphorus compounds. The toxicity of 2-sec-butylphenyl N-methyl carbamate (BPMC) increased by a factor of approximately 15 at the most sensitive time tested. Because the phosphorothioate (P=S) type of organophosphorus insecticides had a synergistic effect on BPMC and the direct-acting organophosphorus insecticide dichlorvos (P=O type) did not, the investigators suggested that inhibition of mixed-function oxidases, which occurs only with the P=S type, is a probable mechanism of this synergism. However, several additional tests of that hypothesis suggested that some other mechanism could also be operative for organophosphate synergism with N-methyl carbamates (Takahashi et al., 1987).

Degrees of Interaction

The greatest departure from dose additivity reported among anticholinesterase insecticides appears to be an increase in malathion toxicity by a factor of about 100 achieved with acute doses and rather unrealistic routes of exposure (Murphy et al., 1959). The first reported example of substantial synergism among anticholinesterase compounds involved binary mixtures of malathion and EPN (Frawley et al., 1957). Both chemicals are anticholinesterase organophosphorothioate insecticides. Tests of the acute toxicity of an equitoxic mixture of the two in rats revealed about a 10-fold increase in toxic mortality. DuBois (1961) reviewed similar but less extensive acute-toxicity tests on dogs that suggested approximately a 50-fold increase. In addition, early feeding studies with EPN at 3 ppm and malathion at 8 ppm in the diet (these were the legal tolerance limits for these chemicals in fruits and vegetables) resulted in increased toxicity, as indicated by erythrocyte cholinesterase inhibition.

As noted earlier, DuBois (1961) tested a large number of binary mixtures of organophosphorus insecticides. He applied the principle of dose additivity—that is, that two compounds with the same mode of action, parallel dose-mortality curves, and similar time and mechanism of action exhibit dose additivity (not synergism) if the simultaneous administration of half the LD_{50} of each results in 50% mortality. Four pairs (malathion and EPN, malathion and dipterex, malathion and Co-Ral, and dipterex and Guthion) resulted in synergism by the dose-additivity definition, as indicated by 100% mortality. Further acute-toxicity tests in rats with a range of doses of equitoxic mixtures of the same pairs resulted in a measure of the degree of synergism when the ratios of the expected LD_{50} of the mixture (if additive) to the observed LD_{50} were calculated. The ratios (degree of synergism) ranged from 1.5:1 (for dipterex and Guthion) to 2.4:1 (for malathion and Co-Ral). The comparison technique probably can be extended to combinations of three or more chemicals, although this does not yet appear to have been done in a published paper.

Using the dose-additive model, McCollister et al. (1959) reported acute toxicity of 50–50 mixtures of the organophosphorus insecticide O,O-dimethyl-O-(2,4,5-trichlorophenyl) phosphorothioate (Ronnel) with each of 10 other organophosphorus insecticides and calculated the ratios of expected (if additive) to observed LD_{50}. Tests of Ronnel with each of six chemicals yielded ratios greater than 1.0:1 (1.3:1 with Systox, 1.4:1 with phosdrin, 1.7:1 with Guthion, 1.8:1 with parathion, 2.1:1 with malathion, and 3.2:1 with EPN); tests with four other pairs yielded ratios of 1.0:1 or less. Of the compounds cited above, only malathion contains carboxyl ester moieties, which are vulnerable to attack by carboxyl esterases, which in turn are known to be sensitive to inhibition by several organophosphorus compounds (DuBois, 1969; Murphy, 1969).

A few other published studies have revealed a slight to moderate (less than 10 times) degree of synergism of acute toxicity of organophosphorus insecticides given simultaneously as binary mixtures to laboratory animals. One criterion that appears to apply to most of the cases of reported synergism is that at least one of the compounds has a higher potency as a carboxyl esterase inhibitor than as an anticholinesterase. DuBois (1961) suggested that the residue-tolerance limits for such compounds should be based on the dosages that inhibit their detoxification enzymes, rather than on the less sensitive acetylcholinesterase inhibition. That suggestion has not, to our knowledge, been adopted as a regulatory rule. From the standpoint of protecting against synergism among organophosphorus compounds, standards for individual compounds based on this detoxification principle might not include any special considerations or extra safety factors (other than the assumption of dose additivity) required for drinking water standards for mixtures containing this class of compounds.

If one considers a case of minimally detectable synergism demonstrated in laboratory animals with a binary mixture of organophosphorus insecticides—i.e., feeding the maximal acceptable dietary-tolerance limits of malathion at 8 ppm and EPN at 3 ppm (Frawley et al., 1957)—one can draw some conclusions regarding the relationship of doses carrying some risk to the dose that might be obtained from drinking water. Assuming ingestion of 1 kg of food, all of which contains maximal food-tolerance limits, a human would ingest 8 mg of malathion and 3 mg of EPN. There are few data available regarding measured concentrations of those compounds in groundwater (or drinking water), but data from California (NRC, 1986) indicate that the highest concentration of malathion observed in groundwater is 23 parts per billion (ppb). If an adult consumed 2 liters of this water, the total dose of malathion would be only 0.046 mg—slightly more than 0.5% of the lowest reported daily amount of malathion for a detectable synergistic effect in chronic dietary feeding tests.

The committee could find no groundwater or drinking water concentration data on EPN. However, for several compounds with similar uses—e.g., parathion, diazinon, and Delnav—the maximal groundwater concentrations reported for California were 4, 9, and 25 ppb, respectively, i.e., no more than 0.050 mg per adult ingesting 2 liters/day, or about 1% or less of the reported minimal dosage of EPN that produced a detectable synergistic effect in animal feeding studies. In fact, Moeller and Rider (1960) tested human response to the dosages that might be obtained at the food-tolerance limits and reported that 3 mg of EPN and 8 mg of malathion in the daily diet of healthy men for 6 weeks led to no observation of depression of plasma or erythrocyte cholinesterase. On the basis of that most-studied example of joint action by organophosphorus insecticides, it appears that no excess risk of cholinesterase inhibition in healthy men is likely if intake of EPN and malathion does not exceed the maximum that could result from legal food residues.

No similar data base is available for other combinations of anticholinesterase organophosphorus or carbamate compounds. There are apparently no reports of tests on interactions that include the carbamate insecticide aldicarb, which has been found in many groundwater samples. In California, maximal concentrations of 47 ppb in groundwater would be equivalent to a 0.094-mg dose in 2 liters of water ingested by adults. Contamination at 47 ppb is more than 4 times EPA's health advisory not to exceed 10 μg/liter (10 ppb) for a 10-kg child. If aldicarb acts in synergy with other anticholinesterases, as Takahashi et al. (1987) have reported for some other carbamates, the risk of occurrence of adverse interactions could be substantially increased.

A related approach based on the dose-additive model is to use the concept of toxic equivalence (Bellin and Barnes, 1987; Eadon et al., 1986). A possible toxic-equivalence scheme for regulation could be used for a mixture of aldicarb and its transformation products aldicarb sulfoxide and aldicarb sulfone. All those compounds are cholinesterase inhibitors, and their potencies relative to that of aldicarb (as measured by 1/NOAEL) can be expressed (Table 5-1). Aldicarb has a lifetime health advisory guideline of 10 μg/liter (EPA, 1987), and the toxic equivalents of aldicarb can be compared with this value. A liter of water containing a mixture of aldicarb at 2 μg/liter, aldicarb sulfoxide at 5 μg/liter, and aldicarb sulfone at 30 μg/liter could be expressed

TABLE 5-1   Relative Potencies and Toxic-Equivalent Concentrations of Aldicarb and Its Transformation Products

Compound	NOAEL, mg/kg	Relative Potency	Concentration, μg/liter	Toxic-Equivalent Concentration, μg/liter
Aldicarb	0.125	1	2	2
Aldicarb sulfoxide	0.125	1	5	5
Aldicarb sulfone	0.6	0.2	30	6

in toxic equivalents of aldicarb on the basis of relative potencies. The toxic-equivalent concentrations for the individual compounds are obtained by multiplying the concentration by the relative potency. For example, for aldicarb sulfone, the concentration of toxic equivalence is

$$(30 \ \mu g \ of \ aldicarb \ sulfone/liter) \ (0.2) \ = \ 6 \ \mu g \ of \ aldicarb/liter.$$

It should also be noted that the toxic equivalence of the mixture is 13 μg/liter—the sum of the concentrations in the last column—which can be compared with the health advisory guideline of 10 μg/liter (EPA, 1987). This approach assumes, as explained in an earlier chapter, that these compounds do not act synergistically.

CONCLUSIONS

The known mechanisms of anticholinesterase synergism depend on interference with or competition for metabolic mechanisms of detoxification of the anticholinesterases or their precursors. Therefore, one might predict that synergism will occur only when the dosage exceeds the theshold where metabolism becomes a rate-limiting factor in toxicity. Of course, that dosage becomes smaller as critical pathways of detoxification are inhibited by other compounds.

Without specific knowledge of the mechanism of synergism and without quantitative data on response to a range of doses of interactive chemicals, it is not possible to determine at precisely what concentrations interaction occurs. From acute-toxicity studies, it appears likely, at least for the compounds discussed in this chapter, that there are dosages below which interactions do not occur and that these can be predicted from data on individual compounds.

With regard to the interaction resulting from the existence of cholinesterase inhibition as a common action, one would anticipate that at most this interaction would result in additive activity. In fact, DuBois (1961) reported that the oxygen analogues of EPN and malathion are strictly additive with respect to their anticholinesterase action in vitro and, in contrast, synergistic in vivo. If the compounds compete for the same active catalytic sites on the acetylcholinesterase molecules, chemicals that are intrinsically less effective as inhibitors might sometimes occupy these sites at the expense of intrinsically more active inhibitors. When that happens, the combined action will be manifested as antagonistic, according to the principles put forward by Veldstra (1956).

RESEARCH RECOMMENDATIONS FOR MIXTURES OF ANTICHOLINESTERASES

- Whether interactions with active inhibitors at a primary biochemical target (i.e., acetylcholinesterase-active center) produce other than additive responses on exposure to multiple chemicals is not known and should be the subject of research. The additivity of multiple compounds at low doses or concentrations should be tested. The resulting knowledge would help to validate the usefulness of a summation or hazard-index approach to recommending quality standards.
- The role of inhibition of carboxyl esterases or other noncritical (silent) receptors in the loss of anticholinesterases, whether or not they involve carboxyl ester linkage, should be investigated further.
- The mechanisms of interaction of carbamate and organophosphorous insecticides recently reported by Takahashi et al. (1987) need better definition to determine whether new concepts or methods for testing interaction potential can be developed.

REFERENCES

Baker, E. L., Jr., M. Warren, M. Zack, R. D. Dobbin, J. W. Miles, S. Miller, L. Alderman, and W. R. Teeters. 1978. Epidemic malathion poisoning in Pakistan malaria workers. Lancet 1(8054):31–34.

Bellin, J. S., and D. G. Barnes. 1987. Interim Procedures for Estimating Risk Associated with Exposures to Mixtures of Chlorinated Dibenzo-*p*-dioxins and Dibenzofuran (CDDs and CDF). U.S. Environmental Protection Agency Report No. EPA/625/3–87/012. Washington, D.C.: Risk Assessment Forum, U.S. Environmental Protection Agency. 27 pp. + appendixes.

Bombinski, T. J., and K. P. DuBois. 1958. Toxicity and mechanism of action of DiSyston. A.M.A. Arch. Ind. Health 17:192–199.

Brodeur, J., and K. P. DuBois. 1964. Studies on the mechanism of acquired tolerance by rats to *O,O* diethyl *S*-[2-(ethylthio)ethyl] phosphorodithioate (Di-Syston). Arch. Int. Pharmacodyn. 149:560–570.

Casida, J. E., R. L. Baron, M. Eto, and J. L. Engel. 1963. Potentiation and neurotoxicity induced by certain organophosphates. Biochem. Pharmacol. 12:73–83.

Cohen, S. D., and S. D. Murphy. 1971. Malathion potentiation and inhibition of hydrolysis of various carboxylic esters by triorthotolyl phosphate (TOTP) in mice. Biochem. Pharmacol. 20:575–587.

Costa, L. G., B. W. Schwab, and S. D. Murphy. 1982a. Differential alterations of cholinergic muscarinic receptors during chronic and acute tolerance to organophosphorus insecticides. Biochem. Pharmacol. 31:3407–3413.

Costa, L. G., B. W. Schwab, and S. D. Murphy. 1982b. Tolerance to anticholinesterase compounds in mammals. Toxicology 25:79–97.

DuBois, K. P. 1961. Potentiation of the toxicity of organophosphorus compounds. Adv. Pest Control Res. 4:117–151.

DuBois, K. P. 1969. Combined effects of pesticides. Can. Med. Assoc. J. 100:173–179.

Eadon, G., L. Kaminsky, J. Silkworth, K. Aldous, D. Hilker, P. O'Keefe, R. Smith, J. Gierthy, J. Hawley, N. Kim, and A. DeCaprio. 1986. Calculation of 2,3,7,8–TCDD equivalent concentrations of complex environmental contaminant mixtures. Environ. Health Perspect. 70:221–227.

EPA (U.S. Environmental Protection Agency). 1987. Aldicarb (Sulfoxide and Sulfone). Health Advisory (Draft). Washington, D.C.: Office of Drinking Water, U.S. Environmental Protection Agency. 16 pp.

Fleisher, J. H., L. W. Harris, C. Prudhomme, and J. Bursel. 1963. Effects of ethyl *p*-nitrophenyl thionobenzene phosphonate (EPN) on the toxicity of isopropyl methyl phosphonofluoridate (GB). J. Pharmacol. Exp. Ther. 139:390–396.

Frawley, J. P., H. N. Fuyat, E. C. Hagan, J. R. Blake, and O. G. Fitzhugh. 1957. Marked potentiation in mammalian toxicity from simultaneous administration of two anticholinesterase compounds. J. Pharmacol. Exp. Ther. 121:96–106.

Lauwerys, R. R., and S. D. Murphy. 1969. Interaction between paraoxon and tri-*o*-tolyl phosphate in rats. Toxicol. Appl. Pharmacol. 14:348–357.

Levine, B. S., and S. D. Murphy. 1977a. Esterase inhibition and reactivation in relation to piperonyl butoxide-phosphorothionate interactions. Toxicol. Appl. Pharmacol. 40:379–391.

Levine, B. S., and S. D. Murphy. 1977b. Effect of piperonyl butoxide on the metabolism of dimethyl and diethyl phosphorothionate insecticides. Toxicol. Appl. Pharmacol. 40:393–406.

McCollister, D. D., F. Oyen, and V. K. Rowe. 1959. Toxicological studies of *O,O*-dimethyl-*O*-(2,4,5-trichlorophenyl) phosphorothionate (Ronnel) in laboratory animals. J. Agric. Food Chem. 7:689.

Moeller, H.C., and J. A. Rider. 1960. Cholinesterase depression by EPN and Malathion. Pharmacologist 2:84.

Murphy, S. D. 1969. Mechanisms of pesticide interactions in vertebrates. Residue Rev. 25:201–222.

Murphy, S. D. 1980. Assessment of the potential for toxic interactions among environmental pollutants. Pp. 277–294 in The Principles and Methods in Modern Toxicology, C. L. Galli, S. D. Murphy, and R. Paoletti, eds. Amsterdam: Elsevier/North Holland.

Murphy, S. D. 1986. Toxic effects of pesticides. Pp. 519–581 in Casarett and Doull's Toxicology: The Basic Science of Poisons, 3rd Ed., J. Doull, C. S. Klassen, and M. O. Amdur, eds. New York: MacMillan.

Murphy, S. D., R. L. Anderson, and K. P. DuBois. 1959. Potentiation of the toxicity of malathion by triorthotolyl phosphate. Proc. Soc. Exp. Biol. Med. 100:483–487.

Murphy, S. D., K. L. Cheever, A. Y. K. Chow, and M. Brewster. 1976. Organophosphate insecticide potentiation by carboxylesterase inhibitors. Proc. Eur. Soc. Tox. XVII, Esc. Med. Int. Cong. 376:292–300.

NRC (National Research Council). 1986. Pesticides and Groundwater Quality: Issues and Problems in Four States. Written by Patrick W. Holden. Washington, D.C.: National Academy Press. 124 pp.

Oonnithan, E. S., and J. E. Casida. 1968. Oxidation of methyl and methyl carbonate insecticide chemicals by microsomal enzymes and anticholinesterase activity of metabolites. J. Agr. Food Chem. 16:28–44.

Polak, R. L., and E. M. Cohen. 1969. The influence of triorthocresylphosphate on the distribution of ^{32}P in the body of the rat after injection of ^{32}P-sarin. Biochem. Pharmacol. 18:813–820.

Roney, P. L., Jr., L. G. Costa, and S. D. Murphy. 1986. Conditioned taste aversion induced by organophosphate compounds in rats. Pharmacol. Biochem. Behav. 24:734–742.

Schwab, B. W., H. Hand, L. G. Costa, and S. D. Murphy. 1981. Reduced muscarinic receptor binding in tissues of rats tolerant to the insecticide disulfoton. Neurotoxicology 2:635–647.

Takahashi, H., A. Kato, E. Yamashita, Y. Naito, S. Tsuda, and Y. Shirasu. 1987. Potentiations of N-methylcarbamate toxicities by organophosphorus insecticides in male mice. Fundam. Appl. Toxicol. 8:139–146.

Veldstra, H. 1956. Synergism and potentiation with special reference to the combination of structural analogues. Pharmacol. Rev. 8:339–387.

6

Volatile Organic Compounds (VOCs): Risk Assessment of Mixtures of Potentially Carcinogenic Chemicals

Drinking water can contain a wide array of toxic substances, including substances known to be carcinogenic or potentially carcinogenic to humans, including benzene, vinyl chloride, carbon tetrachloride, 1,2-dichloroethane, trichloroethylene, chloroform, and other trihalomethanes and other volatile organic chemicals (VOCs). In its approach to mixtures of carcinogens at doses associated with a risk of less than 10^{-3}, EPA (1985) assumes that the upper-bound risk estimates for each of the carcinogenic chemicals can be added. This chapter addresses risk assessment methods for mixtures of low concentrations of carcinogens and draws heavily on a National Research Council (NRC, 1988) report that discussed ways to test the toxicity of complex mixtures and concluded that both exposure and pharmacokinetics are important considerations.

For the known and probable human carcinogens, such as those listed in the paragraph above, maximum contaminant level goals (MCLGs) are set at zero by EPA. Practical considerations might at times make the attainment of zero levels impossible, however, so alternative maximum contaminant levels (MCLs) are set. MCLs are based on technical feasibility and other factors, as well as toxicity. Risk assessment methods can be applied to the individual contaminants to estimate the upper bounds of health risk for alternative hypothesized exposures. The methods used by EPA (1980) apply the linearized multistage (nonthreshold) model.

Table 6-1 lists the MCLs for several compounds. The table also gives estimated lower bounds for drinking water concentrations associated with estimated increases in lifetime risk of developing cancer of 10^{-6}. Exposure at these concentrations is assumed to be constant throughout a lifetime.

TABLE 6-1   Maximum Contaminant Levels and ''Virtually Safe Doses'' for Selected Volatile Organic Chemicals (VOCs) Regulated in Drinking Water

Substance	MCL, mg/liter[a]	VSD, mg/liter[b]
Trichloroethylene	0.005	0.0026
Carbon tetrachloride	0.005	0.00027
1,2-Dichloroethane	0.005	0.00038
Vinyl chloride	0.002	0.000015
Benzene	0.005	0.0012

[a]From EPA, 1987.
[b]Dose associated with lifetime risk of 10^{-6} of developing cancer.

Comparison of these sets of numbers with the measured concentrations in drinking water reported in Table 5-3 shows that the MCL is sometimes exceeded. If the MCL is satisfied, however, the increase in risk of cancer is estimated to be less than 10^{-5} for benzene and trichloroethylene. Because the dose-response curve is typically assumed to be linear at low doses, the risks associated with MCL exposures to carbon tetrachloride and 1,2-dichloroethane are less than 2×10^{-5}. For vinyl chloride, the maximum risk would be about 1.3×10^{-4}.

RISK ASSESSMENT METHODS

Risk assessment is a means to estimate the probability and possible magnitude of a health response associated with a given exposure. For carcinogens, methods of using available data to estimate risk have been relatively well delineated (EPA, 1984, 1986a,b; OSTP, 1984), although validation of the estimates (and of the methods generally) is still seriously incomplete. Animal bioassays are undertaken at rather high doses, where the response (if any) is assumed to be more frequent than at low doses and hence more likely to be observed in a small group of animals. It is then necessary to adopt some mathematical model of the dose-response curve and to extrapolate the response observed at high doses to an estimated response at exposures of interest. The linearized multistage model is widely used to estimate cancer risks associated with environmental exposures (EPA, 1987) and is said to provide an upper-limit estimate of low-dose response. To some degree, the model's wide use reflects its mathematical flexibility. However, biologic support for the assumption of linearity at low doses remains largely inferential and probably wrong in a high proportion of cases (Bailar et al., 1988).

The NRC report on complex mixtures (1988) concluded that, at doses for which the relative risk is less than 1.01 and under the assumptions of the multistage model, the excess risk associated with exposure to several carcinogens can be estimated by adding the excess risks associated with each carcinogen. That report defined a low dose as one associated with an excess risk no more than 1% and with a small relative risk of cancer in the exposed group (no more than 1.01). That result was also inferred to hold for several additional dose-response models used to estimate cancer risks, such as the one-hit, Moolgavkar-Knudsen, linear, and multiplicative models. These inferences are not, however, validated by direct evidence.

Under most circumstances, one would like to use the dose at the target organs as the input to dose-response models. That would increase the precision and could decrease the bias of any risk estimate (Krewski et al. 1987; Whittemore et al., 1986); however, the lack of validated pharmacokinetic models to represent mixtures impedes this effort. Further development of pharmacokinetic models should make possible their application to the individual components of a mixture. For materials activated, say, in the liver or detoxified in the kidney, dose at the liver or the kidney might be most important.

The NRC report (1988) also described risk assessment methods for using data from epidemiologic studies, but noted that the problems presented by human heterogeneity, the potential for bias, the lack of a uniform study design, and the variability in data quality lead to a flood of risk assessment methods, as well as uncertainty about results. Studies based on occupational exposures, which are rarely of known magnitude but can be significantly higher than those of the general population, also present some extrapolation problems. The dose-response models used with these data are often among those for which response additivity is presumed to hold at low doses.

This chapter is related to EPA's existing guidelines for carcinogen risk assessment (EPA, 1986a), inasmuch as the methods suggested here could be applied by EPA in conjuction with those guidelines. Understanding of the biology underlying carcinogenic mechanisms is rapidly evolving and already raises some questions about the need to improve models and assumptions or substitute alternative ones. The adoption of alternative models would necessarily require reexamination of the conclusions given here.

The MCLs for the VOCs and other carcinogens in drinking water are generally below those associated with a 10^{-3} excess risk (see Tables 6-1 and 4-3)—thus they satisfy the first criterion for defining low dose. For doses associated with excess risks greater than 10^{-3}, if the NRC approach is correct, synergism might be important; in such cases, however, the dose of a single carcinogen itself would also be of concern.

It is more complicated to demonstrate that the second criterion for low dose—namely, a low relative risk—has been met. For many of the VOCs that have been studied, evidence of carcinogenicity is largely from animal studies, vinyl chloride and benzene being exceptions. Except for vinyl chloride, excess tumors are generally seen for the hematopoietic or lymphatic systems. For these lymphomas the background cancer risks for humans are about 0.018 over a lifetime (Schneiderman et al., 1979). Hence, for the relative risks to be less than 1.01, the incremental risks associated with exposure to these VOCs must be less than 1.8×10^{-4} to satisfy the low-dose criterion; exposure at the MCLs satisfies the criterion. An argument can be made that there is not necessarily a strict correspondence across sites of cancer for rodents and humans; hence, total cancer rate should be considered. Alternatively, one could consider other individual sites or groups of sites in humans that might be related to sites of observed tumors in rodents. In such an approach, the low-dose criterion is satisfied, because environmental concentrations of the individual VOCs need only be such that the risk of cancer associated with individual toxic compounds is less than 0.003 (0.01×0.33, which is the probability of developing cancer during one's lifetime for all races and both sexes) (Schneiderman et al., 1979).

Vinyl chloride exposures in humans are associated with an excess of liver angiosarcoma, a relatively rare cancer. Background incidence rates are not available for that cancer, but for total liver cancers, the lifetime incidence for both sexes is 0.0054 (J. Horme, National Cancer Institute, personal communication, 1988). If total liver cancers are considered, satisfaction of the low-dose criterion requires that excess risks associated with the chemical in drinking water be less than 5×10^{-5}. If total cancers are considered, however, the second criterion is much more easily satisfied for vinyl chloride. The apparent paradox is due to the smallness of the background incidence of liver angiosarcoma, compared with the total cancer incidence.

The subcommittee recognizes that the assumption of response additivity at low doses does not have an extensive empirical foundation. Rather, it rests on theoretical considerations and observations in limited epidemiologic studies. Regarding higher doses, responses greater than additive occur after human exposure to some mixtures of agents, such as cigarette smoke and asbestos, at concentrations that produce a high incidence of effects separately (NRC, 1988).

CONCLUSIONS AND RECOMMENDATIONS

The acceptance of an underlying dose-response model allows estimates of lifetime cancer risk to be made for individual carcinogens in drinking water. The kind of models also indicate that the risks associated with simultaneous exposure to two or more carcinogens can be added when their concentrations are low, as they are when drinking water standards are satisfied. Hence, in the models advocated in the current EPA guidelines (EPA, 1986a), any potential for more than additivity is ignored in risk assessments of carcinogens present at low doses. Additional research should be conducted to provide a firmer empirical base for those models, so that the techniques for the risk assessment of mixtures can be improved and refined. The present effort is an early step; periodic reevaluation will be necessary.

Any summation of cancer risk estimates must be made with care, because different sets of assumptions can lead to different results. Different models, for example, have been applied to epidemiologic data and animal toxicologic data. In addition, the exposures in epidemiologic studies, although rarely known with any accuracy, are *a fortiori* in the range of past human exposures and often much closer to environmental exposures of current concern than are those of animal studies.

If upper-bound risk estimates are used, even more caveats need to be applied in the summation exercise. It is true that the sum of the upper bounds is an upper bound of the sum, but the sum of the upper 95% confidence limits is not such a limit for the sum. Moreover, the calculations of these upper bounds for the multistage model (EPA, 1986a) are heavily influenced by the linearity of the underlying dose-response data. Relatively linear data give rise to relatively tight confidence limits; nonlinear data generally do not, because data at the higher doses become less and less informative about lower doses as curvature over the intervening range of doses increases.

For carcinogens, the most important research needs are those associated with the development of better estimates of dose-response relationships and risks for individual carcinogens in drinking water. Additional research also needs to be directed toward a better understanding of the mechanisms of carcinogenesis and the development of improved risk assessment models that better reflect the underlying biology. The replacement of existing models with improved models could alter the conclusions about the assumption of response additivity at low doses.

REFERENCES

Bailar, J. C., E. C. C. Crouch, S. Rashid, and D. Spiegelman. 1988. One-hit models of carcinogenesis: Conservative or not? Risk Anal. 8(4):485–497.

EPA (Environmental Protection Agency). 1980. Chloromethane and chlorinated benzenes proposed test role; amendment to proposed health effects standards. Fed. Regist. 45(140):48524–48566.

EPA (U.S. Environmental Protection Agency). 1984. Proposed guidelines for carcinogen risk assessment. Fed. Regist. 49(227):46294–46301.

EPA (U.S. Environmental Protection Agency). 1985. Proposed guidelines for the health risk assessment of chemical mixtures. Fed. Regist. 50(6):1170–1176.

EPA (U.S. Environmental Protection Agency). 1986a. Guidelines for carcinogen risk assessment. Fed. Regist. 51(185):33992.

EPA (U.S. Environmental Protection Agency). 1986b. Guidelines for the health risk assessment of chemical mixtures. Fed. Regist. 51(185):34014–34025.

EPA (U.S. Environmental Protection Agency). 1987. National primary drinking water regulations—Synthetic organic chemicals; monitoring for unregulated contaminants; final rule. Fed. Regist. 52(130):25690–25717.

Krewski, D., D. J. Murdoch, and J. R. Withey. 1987. The application of pharmacokinetic data in carcinogenic risk assessment. Pp. 441–468 in Drinking Water and Health, Vol. 8. Pharmacokinetics in Risk Assessment. Washington, D.C.: National Academy Press.

NRC (National Research Council). 1988. Complex Mixtures: Methods for In Vivo Toxicity Testing. Washington, D.C.: National Academy Press. 227 pp.

OSTP (Office of Science and Technology Policy). 1984. Chemical carcinogens; Notice of review of the science and its associated principles. Fed. Regist. 49(100):21594–21661.

Schneiderman, M. A., P. Decoufle, and C. C. Brown. 1979. Thresholds for environmental cancer: Biologic and statistical considerations. Ann. N.Y. Acad. Sci. 329:92–130.

Whittemore, A. S., S. C. Grosser, and A. Silvers. 1986. Pharmacokinetics in low dose extrapolation using animal cancer data. Fundam. Appl. Toxicol. 7:183–190.

7

Conclusions and Recommendations for Research

In the course of its deliberations, the Subcommittee on Mixtures concluded that assessment of the toxic impact of contaminants in drinking water must include consideration of all routes of exposure to the water, such as inhalation and skin contact, as well as the totality of other exposures to the same contaminants through food, air, and soil. The subcommittee also concluded that, in view of the results shown to date, both response-surface models and physiologically based pharmacokinetic models have potential in the risk assessment of mixtures in drinking water when a small number of materials are to be considered. These techniques offer promise of substantial progress in experiment design and in describing the dose-response relationships associated with exposure to multiple chemicals. The possibility of synergism cannot be ignored; thus, risk assessment approaches that consider only individual toxic agents are inadequate, and approaches based on (dose) additivity are best applied to groups of agents that have the same or similar mechanisms of action at the same biologic sites.

One possible way to incorporate all reported systemic toxicities into a unified measure would be to combine related toxicities and apply weighting factors to the hazard index (and other dose-additive models) currently used by EPA, thus taking into account differences in the toxic spectra of different materials. Current models of carcinogenesis assume no exposure threshold for response, but assume response additivity at low doses, ignoring any potential for synergism. More data are needed to support these models, both to improve estimates of dose-response relationships and risk associated with individual carcinogens in drinking water and to improve understanding of the mechanisms of carcinogenesis.

168

The subcommittee concluded that it would be useful for monitoring purposes to have a simple single chemical analytic method that could produce a measure of the total concentration of members of a group (such as the trihalomethanes or all the volatile organic chemicals) whose risk might be assessed jointly. However, it is questionable whether such a method can be developed to determine the sum of the concentrations of a class of toxicologically similar contaminants in drinking water without potentially confounding detection of other constituents in the water.

The subcommittee proposed the following recommendations for research to improve the assessment of risk associated with exposure to mixtures of chemical contaminants in drinking water. Because of the lack of necessary scientific information, these recommendations are offered as only the beginning of what should be a continuing process. Risk assessment methods related to exposure to multiple chemicals in drinking water and their underlying models and assumptions must be evaluated periodically as more data become available.

1. The usefulness of various indexes used to characterize a mixture of toxicants needs to be studied. For example, can the hazard index be used to predict toxic severity? Specifically:

   a. The hazard index used by EPA should be modified to take into account the sensitivity of each toxic end point. Modifying the hazard index by using a weighting factor would help to account for differences in the toxic spectra of different materials and avoid the lumping together of unrelated toxicities, but would still permit the incorporation of all reported toxicities into a unified measure.

   b. The additive approach should be modified by incorporating a synergism factor that would vary with the amount and type of information available and the concentrations of contaminants.

   c. The "toxic-equivalence" approach that estimates the combined toxicity of the members of a class of chemicals on the basis of the toxicity of a representative of the class should be further developed. Toxicities can then be added according to a weighting procedure.

   d. The usefulness of response-surface methods in describing the relationship between concentration and response for a variety of toxicant combinations should be explored. Relevant experimental designs should be used, and procedures for presenting multidimensional surfaces graphically should be developed.

2. The response additivity and dose additivity of multiple compounds at low doses or low concentrations should be tested. For example, clarification is needed as to whether interactions with active acetylcholinesterase inhibitors at a primary biochemical target (e.g., the acetylcholinesterase-active center) produce truly additive responses when multiple chemicals are used. The

resulting knowledge would help to validate the usefulness of summing responses or doses that produce a given response as an approach to recommending quality standards.

3. The role of inhibition of carboxyl esterases and the role of noncritical (silent) receptors, whether or not they involve carboxyl ester linkage, should be investigated further.

4. The mechanisms of interaction of carbamate and organophosphorus insecticides need better definition to determine whether new concepts or methods for testing for interaction potential can be developed.

5. Biologic research is needed to improve understanding of the mechanisms of cholinesterase inhibition and other actions of the components of mixtures. The results should be used to develop models of these mechanisms; the validity of models should be tested repeatedly as new biologic information becomes available, such as that on the formation and persistence of DNA adducts.

6. Potential pharmacokinetic changes of individual representative chemicals (i.e., chemicals taken to represent classes of chemicals) under the influence of long-term, low-dose intake of a mixture of contaminants in drinking water should be studied.

7. Physiologically based pharmacokinetic models for multiple chemicals in mixtures that represent drinking water contaminants should be developed.

8. Toxicity studies that can help to define the health effects of long-term, low-dose intakes of chemical mixtures in drinking water should be carried out. The assessment of health effects should include immunotoxicologic studies and initiation-promotion studies.

9. The most serious omission in the current literature on mixtures is that of information on dose-response relationships. Some phenomena that prevail at high doses might be absent or attenuated at low doses while others are proportionally strengthened as the statistical models suggest. Some binary combinations are known to produce synergism. It would be reasonable to begin testing the effects of dose by turning to those combinations. Later, observations of the results of largely acute and subchronic exposures could be amplified usefully with longer treatments, inasmuch as environmental problems stem from chronic exposure. Fractional factorial designs could reduce the resource requirements for such tests.

10. The frequency of occurrence of toxic interactions among drinking water contaminants and whether threshold concentrations exist for such interactions should be studied. A computerized data base on toxic interactions would be a useful first step in conducting the needed studies.

APPENDIX **A**

An Approach for Risk Assessment
of Volatile Organic Chemicals in
Drinking Water That Uses
Experimental Inhalation Data and
a Physiologically Based
Pharmacokinetic Model

The toxicity of trichloroethylene appears to be associated with its metabolites, and not with the compound itself (NRC, 1986). Figure A-1 is a plot of AMEFF (the effective concentration of reactive metabolite formed in a compartment of specified volume) against inhalation exposure concentrations obtained from computer simulation of the physiologically based pharmacokinetic model for trichloroethylene. The lowest dose reported in the literature to produce any effect (a hepatic effect) in rats was 56 parts per million (ppm) (NRC, 1986, p. 187). Under the conditions of the experiments, that exposure would produce 940 mg of reactive metabolite in the liver per liter of liver volume (called effective concentration, target concentration, or delivered dose). Interroute extrapolation with physiologically based pharmacokinetics results in the information presented in Figure A-2, where four drinking water consumption patterns are simulated to produce four curves for AMEFF versus drinking water exposure concentrations. As shown in Figure A-2, it takes trichloroethylene at 380–594 mg/liter, depending on the drinking water consumption pattern, to form the AMEFF of 940 mg/liter in the rat liver. Under the assumption that the effective concentration of the toxic metabolite at the cellular or molecular level is the same for all species, further interspecies extrapolation with physiologically based pharmacokinetics yields the information in Figure A-3 for humans. The predicted results suggest, for instance, that the toxic AMEFF would result if a human drank 2 liters of water that contained trichloroethylene at 1,528 mg/liter in six equal portions every day for a lifetime.

171

Other important applications are possible for interroute simulation by physiologically based pharmacokinetic modeling, as described above. For instance, Chapter 2 discussed the issue of exposure to chemicals in contaminated drinking water via multiple routes. In addition to ingestion, inhalation exposures to compounds that volatilize indoors from various uses and cutaneous exposures from bathing and washing are important. Physiologically based pharmacokinetic modeling can be used to estimate the intake of one or more compounds under these circumstances.

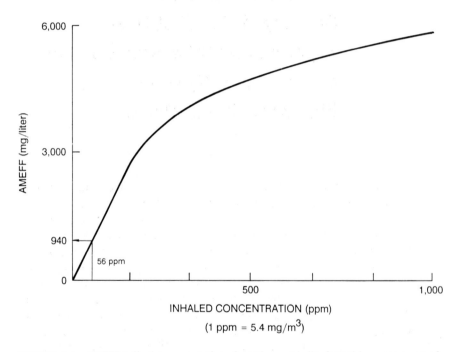

FIGURE A-1 AMEFF (effective concentration of reactive metabolite formed in compartment of specified volume) vs. inhalation-exposure concentration of trichloroethylene in rat. Computer simulation from physiologically based pharmacokinetic model. From NRC, 1986.

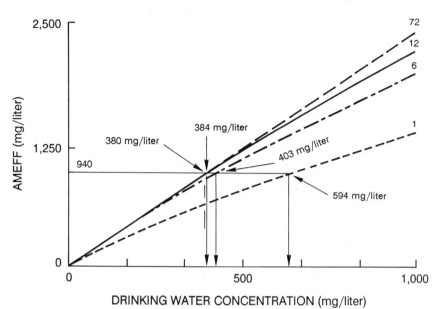

FIGURE A-2   Dose-route extrapolation for trichloroethylene from inhalation exposure of rats to drinking water exposure with four patterns of drinking water intake. Computer simulation from same physiologically based pharmacokinetic model as in Figure A-1. Numbers at ends of drinking water curves are numbers of equal doses (drinking water patterns) taken by test animal. From NRC, 1986.

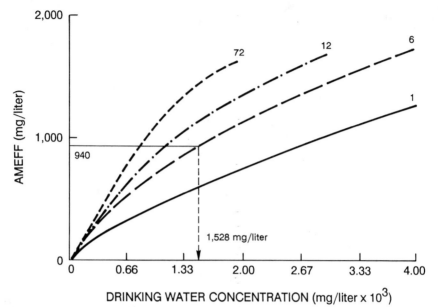

FIGURE A-3   Interspecies (rats to humans) extrapolation for trichloroethylene, based on physio-logically based pharmacokinetic model of equivalent target-tissue doses (AMEFF, 940 mg/liter). Numbers at ends of drinking water curves are numbers of equal doses (drinking water patterns) taken by human. From NRC, 1986.

REFERENCE

NRC (National Research Council). 1986. Drinking Water and Health, Vol. 6. R. Thomas, ed. Washington, D.C.: National Academy Press.

APPENDIX B

A Model Illustrating Synergism

The practical significance of synergism might be considered in a simple mathematical function based on exposure and magnitude of effect.

Nearly all the data available on interactions come from observations based on high experimental or therapeutic doses. Drinking water standards are predicated on low-dose exposures, conditions under which the synergism data might not be duplicated. Although empirical data are lacking, some formal statistical models of the impact of joint exposure suggest that, as the dose (and the consequent effect) is reduced, the contribution of interaction to total toxicity or action is disproportionately attenuated.

A simple mathematical model should help to make clear how this might happen. This model is intended solely to be illustrative. The subcommittee does not advocate the unquestioned use of this model, nor does it suggest that it necessarily reflects the type of response to exposure to a mixture.

Suppose that the magnitudes of toxicity can be described by Equation 1:

$$T = B_0 + B_1 x_1 + B_2 x_2 + B_{12} x_1 x_2, \qquad (1)$$

where T is average toxicity, B_0 is background, B_1 is relative effect of agent 1, B_2 is relative effect of agent 2, B_{12} is "interaction" effect, and x_1 and x_2 are concentrations of agents 1 and 2, respectively.

Consider the following two exposure patterns:

$$\text{High:} \quad x_1 = 5, \qquad x_2 = 10$$

$$\text{Low:} \quad x_1 = 0.5, \qquad x_2 = 1.0.$$

175

Assume that $B_0 = 10$, $B_1 = 7$, $B_2 = 5$, and $B_{12} = 0.3$. Then, the total effect at a high dose with an interaction would be 110 and without an interaction would be 95. At a lower dose, total toxicity with an interaction would be 18.65 and without an interaction would be 18.5.

Thus, at high doses, the interaction makes an important contribution (about 15%) to the toxicity. At low doses, the interaction contribution is only 0.8%.

APPENDIX C

Models of Response: Dose Additivity and Response Additivity

Two definitions of synergy have attained considerable currency, although they are based on distinct and largely incompatible statistical concepts and relate to different biologic views of interaction. This brief discussion is intended only to clarify ideas. It is not a rigorous treatment.

For simplicity, assume that:

a. Only two agents need be considered.
b. Only a single response (yes-no or quantitative) is of interest.
c. The dose-response relationship is monotonic in both agents; that is, an increase in dose of one or the other (or both) is never associated with a decrease in response.
d. Study samples are big enough to ignore random variation.

The relations among the two agents and the single response can be charted like a topographic map. Figure C-1 refers to the proportion (or probability) of animals dead in a yes-no response, but could just as well show (for example) average creatinine clearance or average weight loss.

Note that the response at the origin, where both doses are zero, is what we usually consider "background" and that responses along the two axes (where one or the other dose is zero) yield the usual single-agent dose-response curves: 10% dead at this dose, 20% dead at that dose, etc.

To simplify notation, let $r(a,b)$ designate the response when agent A is given at dose a, and agent B at dose b. Thus, $r(0,0)$ indicates zero dose of both agents and hence designates the background level of response, and $r(a,0)$ and $r(0,b)$ indicate zero doses of one or the other agent and hence the single-agent dose-response curves.

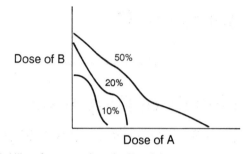

FIGURE C-1   Probability of response (i.e., death) at joint exposure to two materials.

DOSE ADDITIVITY

Pick some point (a,b) in the figure where we are interested in determining whether there is synergy. The response there is $r(a,b)$, and $r(a,b)$ falls on some "topographic contour." Find the ends of the contour and connect them by a straight line. The line can go through $r(a,b)$, or it can be higher or lower (Figures C-2, C-3, and C-4).

If the line goes *through* $r(a,b)$, as in Figure C-3, the agents are said to exhibit dose additivity at that point. For concreteness, if half the LD_{10} of A plus half the LD_{10} of B causes 10% (another LD_{10}) of mortality, A and B are said to exhibit dose additivity in that specific combination.

Note that A and B exhibit dose additivity for *all* combinations that produce LD_{10} (or whatever) if and only if that dose contour is a straight line. Furthermore, A and B show dose additivity over the *whole range* of responses (e.g., all the LD_x) if and only if *every* contour is a straight line. The straight lines need not be parallel, nor need they be equally spaced in any sense, but this is still a very tight restriction.

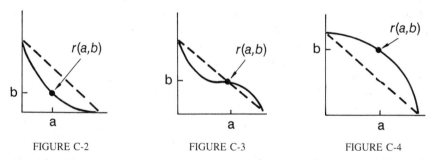

FIGURE C-2   FIGURE C-3   FIGURE C-4

Examples of departure from simple dose additivity upon exposure to two materials. Figure C-2 shows synergism. Figure C-3 shows mixed synergism/antagonism. Figure C-4 shows antagonism.

RESPONSE ADDITIVITY

Another definition of additivity is much closer to that used in other sectors of statistics and mathematics: response at dose a,b is additive if it equals the sum of the separate responses at a and b; $r(a,b) = r(a,0) + r(0,b)$. This definition is often modified for use with dichotomous responses, such as cancer or no cancer and birth defects or no birth defects, in a way that reflects concepts of statistical independence. For these responses, in the notation here and with no allowance for a nonzero background, dose additivity is defined as:

$$r(a,b) = r(a,0) + r(0,b)[1 - r(a,0)]$$

or

$$r(a,b) = r(a,0) + r(0,b) - r(a,0)r(0,b).$$

Allowing for background, $r(0,0)$, by redefining $r(a,0)$, etc., as $r'(a,0) = r(a,0) - r(0,0)/1 - r(0,0)$ gives:

$$r'(a,b) = r'(a,0) + r'(0,b) - r'(a,0)r'(0,b).$$

Of course, if the background rate is nil, $r(0,0)$ drops out, and $1 - r(0,0) = 1.0$, and $r'(a,b) = r(a,b)$ for all a and b.

In the contour graph, draw horizontal and vertical lines from r(a,b) to the two axes, examine the four indicated points, and determine whether the equation above is satisfied:

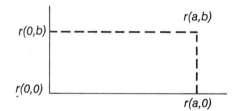

If $r(a,b)$ is too big, A and B are said to be synergistic at that point; if $r(a,b)$ is too small, A and B are antagonistic. It is clear from this graph that the definition of synergy has been profoundly altered. Additivity is now defined in terms of the corners of a rectangle, rather than in terms of an isocontour plus the straight line connecting its endpoints. Response additivity seems to be more tractable in the laboratory (as well as more tractable mathematically), and it is somewhat less restrictive than dose additivity, but these apparent benefits need to be specified more precisely and examined analytically.

Can we integrate the definitions? A and B can be additive in both senses under some limited circumstances, which might be so restrictive as to be of no practical value. In practice, we must choose one or the other.

The model, or definition, that one uses for additivity and for departures from additivity affects the interpretation of experimental data. For example, exposure to two materials from a dose-additive point of view, each at a dose of $LD_{50}/2$, would yield a 50% mortality response. The slope of the dose-response curves is of no consequence in this definition.

Now consider the same mixture from a response-additive point of view. If the actions of the two materials are independent, writing $P(d_1)$ for $r(a,0)$, $P(d_2)$ for $r(0,b)$, and $P(d_1,d_2)$ for $r(a,b)$, the expected result of the combination would be

$$P(d_1,d_2) = P(d_1) + P(d_2)[1 - P(d_1)]$$

or

$$P(d_1,d_2) = P(d_1) + P(d_2) - P(d_1)P(d_2),$$

where $P(d_i)$ is the probability of response at dose d_i ($i = 1,2$). Say further that the two materials have a dose-response curve that is convex upward, with $d_i = LD_{50}/2 = 0.4$. Then response additivity would produce

$$P(d_1,d_2) = 0.4 + 0.4 - (0.4 \times 0.4) = 0.64,$$

rather than the 0.5 expected from a dose-additive model.

By way of contrast, consider two materials with dose-response curves that are concave upward, so that $P(LD_{50}/2) = 0.1$. Response additivity would require that:

$$P(d_1,d_2) = 0.1 + 0.1 - (0.1)(0.1) = 0.19,$$

a result that would be considered much *lower* than the 0.5 anticipated from a doseadditive point of view.

For example, chemicals A and B are tested in various combinations with results as shown below:

		Cancer Incidence Percent at Dose of A, $\mu g/kg$		
		0	10	20
Dose of	0	5%	19%	22%
B, $\mu g/kg$	100	17%	27%	39%
	200	25%	42%	61%

Are the effects of A and B additive, synergistic, or antagonistic with A at 10 g/kg and B at 100 µg/kg?

Response additivity is easily tested inasmuch as $0.22/0.95 = 0.232$ is less than $0.14/0.95 + 0.12/0.95 - (0.14/0.95)(0.12/0.95) = 0.254$. A and B are *antagonistic* at these doses.

To check dose additivity, note that a 27% incidence does not occur on either axis, but (by our assumptions) occurs at some dose higher than 20 g/kg for A and higher than 200 g/kg for B. Thus, the straight line connecting the ends of the 27% contour would be outside (on the far side of the origin from) the point of interest, as in Figure C-2, and A and B are *synergistic* at these doses.

Which definition should be used? Each has substantial and valid uses, and one should not want to proscribe either. What is needed, however, is to make clear which definition is being used in each particular context. As Kodell (1986) points out, "to the pharmacologist and toxicologist, the concept of addition or 'additivity' can imply something about either the doses (concentrations) or the responses (effects) of toxicants acting together. To the biostatistician, addition of doses is in line with the concept of 'similar action,' whereas addition of responses is related to the 'independence' of action. The epidemiologist includes the concept of 'multiplication' of responses . . . that . . . can be interpreted as a type of independence of action.''

REFERENCE

Kodell, R., 1986. Modeling the joint action of toxicants: Basic concepts and approaches. EPA 230–03–87–027 ASA/EPA Conferences on the Interpretation of Environmental Data: Current Assessment of Combined Toxicant Effects, May 5–6, 1986.

Biographical Sketches

SUBCOMMITTEE ON MIXTURES

RONALD E. WYZGA is manager of the Health Studies Program in the Environmental Division at the Electric Power Research Institute in Palo Alto, California. He serves on several committees and chairs several subcommittees of the Environmental Protection Agency Science Advisory Board and the National Research Council. His research interests are environmental risk assessment and health effects of air pollution. Dr. Wyzga received a D.Sc. in biostatistics from Harvard University in 1971. From 1970 to 1975, he worked at the Organization for Economic Cooperation and Development in Paris, where he was a coauthor of a book on economic evaluation of environmental damage. He has also written numerous articles on environmental risk assessment and health effects of pollutants.

JULIAN B. ANDELMAN is professor of water chemistry at the Graduate School of Public Health, University of Pittsburgh, and an associate director of the Center for Environmental Epidemiology, an exploratory research center established in cooperation with the Environmental Protection Agency. He received an A.B. in biochemical sciences from Harvard College in 1952 and a Ph.D. in physical chemistry from Polytechnic University at Brooklyn, New York, in 1960. After postdoctoral work at New York University, he worked for 2 years at the Bell Telephone Research Laboratories in electrochemistry. In 1963, he became a faculty member of the University of Pittsburgh, where his research has centered on the chemistry and public health aspects of water supply systems. He has served on various National Research Council committees, as a consultant to the World Health Organization, on the editorial advisory board of *Environmental Science and Technology*, and on the Drinking Water Subcommittee of the Science Advisory Board of the Environmental Protection Agency.

WALTER H. CARTER, JR., is chairman of the Department of Biostatistics at Virginia Commonwealth University. His principal interest is in the design and analysis of response-surface experiments. For the last 17 years, he has been developing and using those experiments to study the effects of combinations of drugs and other chemicals. He is a coauthor of the monograph *Regression Analyses of Survival Data in Cancer Chemotherapy*. Dr. Carter is a fellow of the American Statistical Association and a member of the Biometric Society.

NANCY K. KIM is director of the Division of Environmental Health Assessment of the New York State Department of Health. She received a B.A. and a Ph.D. in chemistry from the University of Delaware and Northwestern University in 1966 and 1969, respectively. Before Dr. Kim's present appointment, she served as director of the Bureau of Toxic Substance Assessment and as a research scientist in the Division of Laboratories and Research, both in the Department of Health. Dr. Kim is a member of the Environmental Protection Agency Science Advisory Board's Environmental Health Committee and Drinking Water Subcommittee. Her primary professional interests are in environmental health, specifically environmental toxicology, exposure assessment, and risk assessment.

SHELDON D. MURPHY has been professor and chairman of the Department of Environmental Health, School of Public Health and Community Medicine, at the University of Washington in Seattle since 1983. Previous faculty positions were at Harvard University and the University of Texas. Dr. Murphy received a Ph.D. in pharmacology from the University of Chicago in 1958. He has been president of the Society of Toxicology, has held other national and international positions, and has served on expert committees for the State of Massachusetts and the World Health Organization, on the Environmental Health Advisory Board of the Environmental Protection Agency, and on the Board of Scientific Counselors of the National Institute of Environmental Health Sciences and the National Institute for Occupational Safety and Health. Dr. Murphy has written some 200 book and journal articles on the toxicology of pesticides and other chemicals.

BERNARD WEISS received a Ph.D. in 1953 from the University of Rochester, where he is professor of toxicology and deputy director of the Environmental Health Sciences Center. He has been a member of the toxicology study section of the National Institutes of Health, has served on the U.S.–U.S.S.R. Joint Environmental Health Programs and various National Research Council committees concerned with environmental health issues, and has held consultantships to federal agencies. His current service includes the Environmental Protection Agency Science Advisory Board, the National Institute of Environmental Health Sciences Board of Scientific Counselors, and the National Research Council Committee on Neurotoxicology and Risk Assessment. His research activities embrace various aspects of behavioral toxicology and pharmacology, including the actions of such heavy metals as lead, mercury, and manganese, of organic solvents, of central nervous system drugs, and of other substances whose actions are reflected in behavior.

RAYMOND S. H. YANG is a senior staff member of the Carcinogenesis and Toxicology Evaluation Branch of the National Institute of Environmental Health Sciences/National Toxicology Program and an adjunct professor of toxicology at North Carolina State University. His research interests cover many subdisciplines in toxicology, including carcinogenesis, reproductive toxicology, physiologically based pharmacokinetics, toxicology of chemical mixtures, and toxic interactions. Among his current responsibilities is service as principal investigator in a program on the toxicology of chemical mixtures derived from groundwater contaminants, which is related to the Superfund hazardous-waste site cleanup efforts. He is a member of the Society of Toxicology and the editorial board of *Fundamental and Applied Toxicology*.

PART III
Cumulative Index

Drinking Water and Health
Volumes 1–9

Cumulative Index

A

AADI (adjusted acceptable daily intake), (6)171

2-AAF (2-acetylaminofluorene), (6)145

Abbreviations, definitions of, (6)218–219

Abortions, spontaneous, (see spontaneous abortions), (6)

ABS, (see acrylonitrile-butadiene-styrene), (4)

Absorption, (1); (4); (6); (8); (9)
of chemical agents, (1)29
extrapolation of, (8)139–140
gastrointestinal, (8)122
measurement by DNA adducts, (9)7
process in GI tract, (6)210–211
rate of, (8)121
rates, differences between species, (1)32, 53
skin, (8)122–123
systemic, (6)257
(see metabolism), (4)

Acanthamoeba species, (1)113

Acceptable daily intake (ADI), (3)2, 25; (5); (6)171, 254, 257, 296, 410–411
aldicarb, (5)12; (6)309
arsenic, (5)123
carcinogens, considerations of, (3)36–37
chronic exposure, (3)29–37
di(2-ethylhexyl) phthalate, (6)358, 410
dose-response methodology, (3)31–37
nitrofen, (6)379

no-adverse-effect level, (3)31
p-dichlorobenzene, (5)27
trichlorfon, (6)408
uncertainty factors, (3)36

Acceptable risk, (1)24

Acculturation and hypertension, (1)407–409

Acetaldehyde, (1)686–687; (6)50; (7)152–153
effects on animals, (1)687
effects on man, (1)686

Acetic acid (AA), (7)134–135

Acetone, chloroform precursors, (2)158

Acetonitrile, (4)202–206; (7)43
health effects, (4)204–205
metabolism, (4)203–204
SNARL, (4)205–206
TWA standard, (4)203
(see also nitriles), (4)

2-acetylaminofluorene (2-AAF), (6)145; (9)26, 27

Acetylcholinesterase, (6)306; (9)
inhibitors, (9)146–159, 170

Acetylsalicylic acid toxicity, (1)37

ACGIH, (see American Conference of Governmental Industrial Hygienists), (1); (3); (6)105

Acicular crystals, (1)144–147, 158–159

Acidity, high pH conditions, (2)86

Acinetobacter, (2)309

Acneiform skin eruptions, iodine effects, (3)306

187

U

X

Y

Z